THE PROFESSOR, THE INSTITUTE, AND DNA

OSWALD THEODORE AVERY

THE PROFESSOR,
THE INSTITUTE,
AND DNA

BY RENÉ J. DUBOS

THE ROCKEFELLER UNIVERSITY PRESS

NEW YORK · 1976

CONTENTS

ACKNOWLEDGMENTS

I would like to express my gratitude to the following colleagues, both in and out of The Rockefeller University, who read all or part of the manuscript during its preparation and who offered suggestions on various points or shared their memories of O. T. Avery: Saul Benison, Stuart Elliott, Walther Goebel, Michael Heidelberger, James Hirsch, Maclyn McCarty, Walsh McDermott, Robert Morison, Arnold Ravin, and Norton Zinder. Joseph Ernst, Director of the Rockefeller Archive Center, and Sonia W. Mirsky, Associate Librarian, The Rockefeller University, were most generous in helping me locate obscure source material.

THE PROFESSOR, THE INSTITUTE, AND DNA

INTRODUCTION

This book has two heroes: Oswald Theodore Avery (1877–1955) and The Rockefeller Institute (1901–1955). I have had such close associations with both of them that the objective description of facts and events concerning them has often seemed to me less compelling than the subjective remembrance of things past.

I met Avery in 1927 and worked in a laboratory adjacent to his own for the following 14 years. Our relations were so personal that he acted as witness to my marriage in October, 1946, five years after I had left his department. I have been continuously associated with The Rockefeller Institute (now The Rockefeller University) since 1927, except for the years 1942–1944, which I spent at Harvard University Medical School. Since my retirement in 1971, I have continued to occupy the office in which I worked as a member of the scientific staff. There is no place in the world where I have spent as much time as on the Rockefeller campus, and where I feel more at ease. Whenever I approach the stalwart plane trees of the 66th Street entrance, I know "this is the place."

Many of the statements I shall make concerning Avery and the Institute are not based on documents, but on personal observations and memories. Whenever possible, I have checked their accuracy with the few surviving friends and colleagues who, directly or indirectly, participated in the experiences I report. It is obvious, however, that the very nature of my relationship with the two heroes of this book colors my account of them, perhaps at times to the point of distortion. I have tried to acknowledge this difficulty by reporting in the chapter entitled "As I Remember Him" my *interpretations* of Avery's attitudes as I perceived them during the years I worked in his laboratory.

Documents concerning the history of The Rockefeller Institute are available in the archives of The Rockefeller University and of The American Philosophical Society. I have consulted only a few of these primary documents, and have derived most of my information from semiofficial secondary sources and from persons who have been directly involved in the Institute's affairs.

Some limited documentation concerning Avery has been deposited in The Rockefeller University archives and in the Manuscripts Section, Tennessee State Library and Archives in Nashville. I have received much additional information concerning his familial background and his private life from his sister-in-law, Mrs. Catherine Avery (Mrs. Roy C. Avery), from a few of his friends, from Mr. Howard Williams, archivist of Colgate University, and from Dr. Joseph Ernst, Director of the Rockefeller Archive Center.

The development of Avery's scientific career can be followed, of course, from his published papers, but more precisely and interestingly from the detailed annual reports he submitted to the Board of Scientific Directors of The Rockefeller Institute, as well as from reports to the Trustees of The Institute, submitted by the Director of The Rockefeller Hospital. I have quoted extensively from these documents, which are available in the archives of The Rockefeller University.

During my two years at Harvard Medical School, I wrote a book entitled *The Bacterial Cell* (1945), which was profoundly influenced by my earlier associations with Avery. I shall paraphrase below a few lines from the preface to that book, because their spirit is as appropriate today as it was three decades ago.

Those who have been connected with The Rockefeller Institute at some time between 1920 and 1950, will undoubtedly recognize in the following pages echoes of conversations held in the Institute lunchroom and especially in the Department of Respiratory Diseases. I shall be rewarded for my efforts if my account helps them to recapture, and others to imagine, the vital atmosphere of the Institute, and especially the smiling wisdom of one whom we called with admiration, gratitude, and love "The Professor" or, more familiarly, "Fess" Avery.

CHAPTER ONE

THE PROFESSOR
AND THE INSTITUTE

A Dynamic Institution

From the windows of my office in the Bronk Laboratory building of The Rockefeller University, I can see on my right, looking north, the four buildings that constituted the original Rockefeller Institute for Medical Research. The one nearest to me is the Hospital, where Oswald T. Avery's department was located on the sixth floor.

These four buildings were erected between 1906 and 1938, and were designed primarily as laboratories. Even in the Hospital, approximately half of the floor space was assigned to laboratory work. The architectural simplicity and uniformity of the initial Institute ensemble symbolize the singular unity of purpose that presided over its creation—the conduct of laboratory research focused on medical problems.

Several new buildings have been added since the Institute was metamorphosed into The Rockefeller University, and the grounds have been arranged into a formal, parklike campus, the elegance of which calls to mind an Ivy-League atmosphere. The new buildings are more diversified than the old ones, and differ from them greatly in architectural style. The various styles correspond not only to different periods, but, more importantly, to different types of functions, many of which were either nonexistent in the old Institute, or little developed. In addition to the new laboratory buildings, a variety of structures now serve as residences for students, staff, and visitors; as halls for lectures, conferences, concerts, and purely social gatherings; as offices for the administrative requirements of modern academe and for its complex social relationships.

The present character of the campus was determined in part by the transformation of the medical research Institute into an educational institution. It reflects even more, however, changes that have occurred in science and in society during recent decades. Most of the medical research institutes that were created in different parts of the world at the turn of the century have retained their original character, and a few have gone out of existence. In contrast, The Rockefeller Institute has continuously enlarged the scope of its research fields and has undergone profound changes in its physical and administrative structure — to the point of becoming a post-graduate university in which medical sciences are only a part of a much broader academic program. The reason for this continued vigor and ability for self-renewal is certainly to be found in the initial policies that were formulated for the Institute; they were so broad that they enabled it to evolve rapidly by adapting to new scientific trends and new social demands. Some aspects of this adaptability will be considered in Chapter Two. Nowhere in this book, however, shall I have occasion to discuss the University phase of the institution, because it began only in 1955, the very year of Avery's death. In fact, I shall focus my interest on The Rockefeller Institute for Medical Research, which came to an official end in 1953. The qualification "for medical research" was dropped from the name five years after Avery left the Institute for his final retirement in Nashville.

In this introductory chapter, I shall outline what could be readily seen and learned of Avery and of The Rockefeller Institute for Medical Research by an outsider or by a newcomer to the staff, as I was in 1927.

Workshops of Science

At the time of Avery's birth in 1877, Louis Pasteur in France and Robert Koch in Germany were in the process of demonstrating that bacteria and other microorganisms can cause disease in animals and human beings. Their findings had immediate practical applications in the control of disease, and also had the broader social consequence of making the medical and general public understand that progress in the practice of medicine could be greatly accelerated by laboratory investigations that did not involve the care of patients. Obvious as this view has now become, it appeared far-fetched a century ago.

Interest in laboratory science spread so wide and so fast at the end of the nineteenth century that it led to the creation of several medical research institutes where scientists could devote all their efforts to the acquisition of theoretical and practical knowledge. This new trend enabled Avery to abandon clinical medicine at the age of 30 and to opt for a life of scientific research, first at the Hoagland Laboratory in Brooklyn and then at The Rockefeller Institute in Manhattan.

Details concerning the emergence of scientific medicine and the different phases of Avery's life will be presented in subsequent chapters. The emphasis here will be on those aspects of the Institute that made it an environment ideally suited to Avery's life and to his work.

William Blake's phrase "What is now proved was once only imagin'd"[1] could well be applied to the medical research institutes created around 1900, because these were the incarnation of ideals formulated by Francis Bacon and René Descartes in the seventeenth century, at the very beginning of experimental science. In his book, *The New Organon* (1620), Francis Bacon described a utopian scientific community that he called Salomon's House, in which scholars devoted themselves to the search for knowledge "for the benefit and use of life." The ultimate goal of the experiments carried out in Salomon's House was the improvement of man's estate, but Bacon recognized that not all experiments could be expected to lead immediately to practical results. In his words, "Scientists should be willing to carry out a variety of experiments, which are of no use in themselves but simply serve to discover causes and axioms; which I call *experimenta lucifera,* experiments of light to distinguish them from those which I call *fructifera*, experiments of fruit."[2] Bacon's emphasis on the importance of *experimenta lucifera* provided the new research institutes with their operational philosophy: the cultivation of theoretical science as an essential step in the development of practical knowledge.

Descartes also contributed to this operational philosophy by affirming that the best way to foster the advancement of knowledge was to provide scientists not only with material facilities, but also with leisure, peace of mind, and complete freedom. The view that scientists had a right to leisure, even though they were supported by public funds, was truly a new social concept.

Two centuries later, Pasteur restated in memorable phrases Bacon's dream of a Salomon's House and Descartes' plea for intellectual freedom for scientists. Speaking of "these sacred institutions that we designate by the expressive name of laboratories," he urged that they be multiplied and well supported because they are "the temples of wealth and of the future . . . where humanity learns to read in the works of nature." He evoked the happiness that he had experienced "in the serene peace of laboratories and libraries."[3] The Rockefeller Institute for Medical Research was created in 1901 by Mr. John D. Rockefeller, Sr., to provide for scientists such an atmosphere of peace and serenity.

During the early planning stages, there was a widespread belief that the Institute should be linked to some well-established university, medical school, or public health laboratory. However, this plan was vigorously rejected by Mr. Rockefeller himself, for reasons that were strangely reminiscent of the opinions expressed by Bacon and Descartes in the seventeenth century. Mr. Rockefeller feared that clinical duties, the preparation of lectures, the conduct of examinations, and especially the administrative responsibilities associated with medical and educational practices would distract investigators from their research.[4] Unquestionably, other reasons, such as matters of prestige and problems of funding, also played a part in the final decision that The Rockefeller Institute should be completely independent from traditional academic or medical institutions, but an important factor in the decision was Mr. Rockefeller's desire that the Institute investigators have complete intellectual freedom and be protected from extraneous pressures, whether academic or administrative.

The building site that was selected for the new Institute was quite remote from what was then the center of New York City, as if to symbolize the decision of its founders to make it intellectually independent of established centers of medical research. It was situated between 64th and 68th Streets along the East River, and formed on its eastern half a rocky bluff about 40 feet high overlooking the river. The property was still farmland when it was bought in 1901; goats were browsing on the gentle slopes toward its western boundary, now occupied by York Avenue.

The instructions to the Boston firm of architects that was employed for the construction of the Institute were that the style of the buildings should be *"as simple* as is consistent with present purpose, future additions, and *general utility"* (italics mine).[5] Just as the site selected for the Institute was removed from the hustle and bustle of the city and from traditional academic and medical influences, so were the buildings devoid of any pretense to be anything other than places designed for work and thought.

Ground was broken for the first laboratory building, now called Founder's Hall, in July, 1904. When the building was dedicated along with an animal house and a powerhouse on May 11, 1906, not a word was said about its architectural style "either in praise by the visitors, or vaunting by their hosts."[6] The reason for this tactful silence was that the main building was far from sumptuous. It was large and well equipped by contemporary criteria, but looked rather drab, especially in comparison with the new buildings faced with white marble that had just been completed for the Harvard Medical School in Boston.

In contrast to the lack of popular interest in the architecture of The Rockefeller Institute buildings, there was much excitement at the time about the skyscrapers that were being erected in downtown Manhattan — especially about the Woolworth Building, which came to dominate the Manhattan skyline in 1910. The Gothic frills that ornamented that skyscraper from top to bottom made it famous throughout the world as the "cathedral of commerce"—an international shrine to the gods of money and technology.

Until 1957, none of the buildings that were erected on The Rockefeller Institute grounds along East River was influenced by the modern architectural styles. Each retained the same low profile and the same uniformly prosaic institutional appearance that had been adopted for Founder's Hall. The builders must have been instructed to use bricks that were neither red, nor white, nor yellow in color, but nondescript. The buildings clearly were not meant to be cathedrals of science, as the Woolworth Building pretended to be a cathedral of commerce, but rather functional, unpretentious workshops, designed for the prosecution of laboratory research.

The Rockefeller Institute Hospital, in particular, has an austere, functional simplicity that makes it remarkably inconspicuous. It is not sufficiently vast or high to be overpowering or physically inspiring; it is not sufficiently small or cozy to give it an obvious emotional appeal. Despite its outward simplicity, however, it was, at the time of its dedication on October 17, 1910, a highly efficient structure, well suited to the methods then known for the treatment of the sick and for scientific research on disease.

The two thousand visitors who attended that dedication were somewhat surprised, and probably many of them disappointed, to find that the architects and builders had put up "a strictly utilitarian structure . . . space and expenditure for artistic effect being strictly limited by the Directors."[7] These are the very words of T. Mitchell Prudden, one of the initial members of The Rockefeller Institute Board, who had taken a special interest in the planning of the Hospital and of its activities. The visitors were impressed, however, by the efficiency of the wards and of the diagnostic services, by the diet kitchen, which was very unusual for the time in its completeness and relative size, and especially by the importance of the laboratories, with space and equipment far in excess of needs for mere routine examinations and tests.

Today, the original buildings of The Rockefeller Institute for Medical Research look much as they did at the beginning of the century, except for the mellowness that they have acquired from the ivy that covers their walls and from the greenery that surrounds them. They were so soundly built that they have proved adaptable to the changes in laboratory procedures that have continuously transformed medical research during the past few decades. I like to believe that they will serve for many more decades as research laboratories and as shelters for scholarly thought. It is now almost

50 years since I first worked in them, and I still marvel at the quality of their materials and at the soundness of the structure; I still make it a point to walk up and down their broad stairways for the sheer enjoyment of physical contact with their sturdy oak railings and their broad marble steps.

These old buildings call to mind the venerable institutions built in Europe during earlier centuries, when good workmanship assured a quality that transcended fashions and that improved with time. For example, the Royal Institution of London is not remarkable for its architectural style, but it was so well built that it has aged well and has become more appealing with each generation. It still conveys the atmosphere of integrity it had when Michael Faraday worked there until the end of his life. One century after Faraday entered the Royal Institution, Oswald T. Avery joined the Hospital of The Rockefeller Institute, where he stayed for the rest of his professional life.

The Professor and the Genius Loci

When The Rockefeller Institute Hospital opened its doors, Avery had been working for several years on bacteriological and immunological problems at the Hoagland Laboratory in Brooklyn. He was a physician, but he already knew that he was more interested in laboratory investigations than in clinical work. The position offered to him at The Rockefeller Hospital in 1913 did not involve taking care of patients; instead, he was expected to participate as a bacteriologist and immunologist in the laboratory program on lobar pneumonia. He was technically well equipped for this task and, more interestingly, he was admirably suited by temperament to the intellectual and human atmosphere that he found in The Rockefeller Institute.

Just as the planners of the Institute buildings scorned architectural glamor, so did Avery shy away from public performances during his adult life; everything about his person was in a low key that made him inconspicuous, like the buildings in which he worked and lived.

He was small and slender, and probably never weighed more than 100 pounds. In behavior he was low-voiced, mild-mannered, and seemingly shy. His shirts, suits, neckties, and shoes were always impeccable, but were as subdued as his physical person. His demeanor was charmingly courteous, but in a conservative way that often called to mind a buttoned-up *petit bourgeois*. I shall evoke in other chapters the richer and more unusual aspects of his personality, but shall emphasize here the parallelism of his scientific evolution with that of The Rockefeller Institute.

For both Avery and the Institute, the point of departure had been the awareness that the scientific basis of medicine was extremely weak and the belief that the control of disease could be made more rational by knowledge derived from laboratory investigations. In both cases, however, the study of disease led to problems of a nonclinical character, especially having to do with the chemical mechanisms of life processes. Instead of being exclusively concerned with medical research, narrowly conceived, the Institute became more and more chemically oriented. In a similar way, Avery, who started with the study of lobar pneumonia, rapidly moved toward the study of the chemical basis of biological specificity; he ended with the demonstration that hereditary characteristics are transmitted by molecules of deoxyribonucleic acid (DNA), his most celebrated achievement. By integrating his medical training with sophisticated laboratory disciplines, he was a perfect representative of the intellectual attitude that gave its shape to medicine during the first half of the twentieth century. An outline of his scientific contributions is presented in Chapter Six. Technical details are described and discussed in Chapters Seven through Eleven.

Thus, Avery and the Institute had much in common, because they were, respectively, the human and institutional expressions of the same scientific attitudes. They both emerged and developed in the atmosphere of expectancy generated by a few triumphs of scientific medicine at the end of the nineteenth century; both followed an intellectual course that led them from the study of specific diseases to large problems of theoretical biology; both became part of a culture in which laboratory scientists were regarded as members of a kind of priesthood, willing to accept social constraints for the sake of intellectual privileges.

An Avery Memorial Gateway to the Rockefeller campus was dedicated on September 29, 1965. Its great piers, made of red Laurentian granite quarried in Avery's native Canada, bear the simple inscription:

IN MEMORY OF

OSWALD THEODORE AVERY

1877–1955

A MEMBER OF THE FACULTY OF

THE ROCKEFELLER INSTITUTE

1913–1948

ERECTED BY GRATEFUL FRIENDS AND COLLEAGUES

The Gateway is low-key, but bold in design, true to Avery's character.

It is the only entrance to the campus that has been given a name, an indication of the uniqueness with which Avery represented the scientific and social concepts that led to the creation of The Rockefeller Institute. As I remember him, ardently involved in laboratory work, gently but intensely discussing science with collaborators and friends, brooding at his desk, or slowly walking on the grounds in a meditative mood, he symbolizes for me the *genius loci* of The Rockefeller Institute for Medical Research.

CHAPTER TWO

FROM THE BEDSIDE
TO THE LABORATORY

The Rise of Scientific Medicine in the United States

When Avery entered medical school in 1900, the most influential physician in the United States was William Osler, who was then professor of medicine at The Johns Hopkins University. Osler attributed his phenomenal success as a healer to the confidence he inspired in his patients through psychological traits that were unrelated to his scientific knowledge of disease. More generally, he believed that one of the most important aspects of medical care was, in his words, "The Faith that Heals . . . Faith in the gods or in the saints cures one, faith in little pills another, hypnotic suggestion a third, faith in a common doctor a fourth." In his lectures and writings, he emphasized time and time again the therapeutic effectiveness of what he called "psychical methods of cure" or, more simply, faith healing. What he really meant by these expressions is the effect of the psychological influences through which physicians help the automatic process of self-healing in their patients.

The original text of this book included some 10 pages devoted to the place of the various practices of faith healing (self healing) in medical history. However, the four persons who read the typescript felt that this subject should be deleted because it had no "obvious relevance," either to Avery or to The Rockefeller Institute. I have reluctantly followed their advice and shall publish these pages elsewhere, but I must at least state my opinion that, although the relevance of the psychological aspects of healing to scientific medicine is not obvious, it is nevertheless extremely important, and may become even more so in the near future. Early in this century, in fact, this importance was explicitly recognized by William Henry Welch, Simon Flexner, and Walter B. Cannon — physicians who cannot be suspected of antiscience bias, since they were among the chief architects of scientific medicine in the United States, and at The Rockefeller Institute.

I know from conversations with Avery that he, too, was much impressed by the influence of the mind on the phenomena of disease. However, the mystical and irrational character of most faith-healing practices was uncongenial to him, and he felt more at ease with medical problems that could be studied in the laboratory by physicochemical methods. The spectacular achievements of modern medicine are testimony to the effectiveness of this orthodox scientific approach. On the other hand, the proliferation in our times of ways of healing that have no rational basis in the conventional natural sciences strongly suggests that medicine will not become fully scientific until it has come to grips with what Osler called "the faith that heals." A medicine based exclusively on the body-machine concept of human nature may soon be as obsolete as is now the gold-headed cane of the nineteenth-century European physician.

There was little doubt, however, about the direction that medical sciences should take at the turn of the century. The most important medical problems of the time involved infectious and deficiency diseases that could not be significantly influenced by any form of faith healing or self healing, but could be studied effectively by laboratory methods. However, although this kind of experimental medicine had flourished in Europe since the beginning of the nineteenth century, it was practically nonexistent in the United States.

The general view among American physicians was that laboratory science could never contribute anything of practical value to the practice of medicine. Some American hospitals had modest laboratories, but these were only for diagnostic work. Neither the universities nor the medical schools nor governmental bodies were inclined to provide facilities or personnel for medical research.

There were, of course, a few exceptions. A laboratory for experimental medicine — the first in America — had been established at Harvard Medical School in 1871 for the professor of physiology, Henry P. Bowditch, but it consisted of only two small rooms in an attic. Furthermore, Dr. Henry J. Bigelow, who was then the leading spirit of Harvard medicine, warned that it would be dangerous to let students be distracted from useful knowledge by theoretically interesting, but practically useless, learning. "The excellence of the practitioner depends far more upon good judgment than upon great learning," Bigelow wrote. "We justly honor the patient and learned worker in the remote and exact sciences, but should not for that reason encourage the medical student to *while away his time* in the labyrinths of Chemistry and Physiology, when he ought to be learning the difference between hernia and hydrocele"[1] (italics mine). In practice, the student of medicine learned to take care of the sick by serving as an apprentice to an experienced doctor, and the only worthwhile form of medical science was assumed to be the knowledge acquired by observation at the bedside.

The absence of research laboratories did not mean, however, that all of nineteenth-century American medicine was backward, as Frederick Gates erroneously assumed when he conceived the idea of the Rockefeller Institute in 1901 (see page 20). As early as 1765, Dr. John Morgan of Philadelphia had founded the first American medical school as part of Benjamin Franklin College, and had based its curriculum on the scientific experience he had gained in Europe.[2] European medical books were rapidly translated for the American market.

Moreover, the science and practice of medicine had been advanced by several American achievements of great importance — in surgery, in clinical diagnosis, in general anesthesiology, and in descriptive epidemiology. In 1809, for example, Ephraim McDowell performed the first ovariotomy on record anywhere in the world. Between 1822 and 1839, William Beaumont took advantage of the large gastric fistula in his patient Alexis St. Martin to conduct fundamental experiments on digestion. In 1837, William W. Gerhard differentiated typhus from typhoid fever. Between 1842 and 1846, both Crawford W. Long and William Thomas Green Morton contributed independently to the demonstration that ether is highly effective for surgical anesthesia. Daniel Drake made elaborate epidemiological observations that he summarized in *Diseases of the Interior Valley of North America,* published in 1850. Important as these contributions were from the practical point of view, however, they were not part of what came to be known as scientific medicine because they had not required physicochemical understanding of pathological processes.

Here and there in North America, a few nineteenth-century physicians began to investigate the causation and mechanism of disease by trying to interpret careful observations made on patients in the light of what could be learned from the study of post-mortem specimens. The most illustrious representative of this attitude was William Osler. Born and trained in Canada, then Professor of Medicine in Philadelphia and Baltimore, Osler finally moved to England, where he became Regius Professor of Medicine at Oxford University.[3] He was knighted in 1911 and is now remembered as Sir William Osler.

Osler had travelled in Europe and was familiar with the developments in laboratory research. To the end of his life, however, he remained unshaken in his belief that medicine can be learned only at the bedside, and that its most important aspect is the art of establishing the right kind of personal rapport between physician and patient. His prodigious and lasting fame as a clinician is evidenced by the fact that, 20 years after his death, he was referred to as "that quasi-divinity of ours." [4] Yet, the most important advances of modern medicine emerged not from the kind of clinical and pathological observations that he advocated, but rather from laboratory investigations.

Throughout the nineteenth century, a number of young American physicians had spent a few months, or even years, in Europe to familiarize themselves with the new kind of medical science that was then flourishing in the medical centers of Great Britain, France, Austria, and Germany. Several of them contributed to the new learning through their own research efforts while in Europe, but they discovered after their return home that their native land offered no opportunities for laboratory research. Referring to the bright young American physicians who had worked in his laboratory, the German physiologist Karl Ludwig wondered why they were never heard from again, even though they had done brilliant research work while in Germany.[5] Although otherwise typical, the particular case of William Henry Welch has the special merit of referring to the same Welch who eventually became the mastermind of laboratory medicine in the United States—first at The Johns Hopkins Hospital and Medical School, then at The Rockefeller Institute for Medical Research.[6]

In the fall of 1872, Welch entered the College of Physicians and Surgeons, then the best of the three medical schools in New York City. After his graduation in February, 1875, he served an internship at Bellevue Hospital. There he gained some feeling for medical research from the pathologist Francis Delafield, who had imbibed the teachings of the French and German schools. In April, 1876, Welch sailed for Germany to get the feel of the current situation in science and to try his hand at research. His two years abroad were immensely successful from the scientific point of view. He learned much from some of the most famous German pathologists and carried out creditable research in several areas of pathology.

When he returned to the United States in February, 1878, he discovered that the only laboratory positions available anywhere in the country were for teaching elementary microscopy and pathology, without provision for research of any sort. In a letter to his sister after his return home, he discussed his scientific interests, but had to report that there was "no opportunity for, nor appreciation of, no demand for that kind of work here. . . . I sometimes feel rather blue when I look ahead and see that I am not going to be able to realize my aspirations in life."[7] A similar discouraging situation was experienced by other young American medical men returning from Europe—for example, by T. M. Prudden, who was to become, two decades later, part of the group that formulated the concept of The Rockefeller Institute for Medical Research and an influential member of its Board of Scientific Directors. The only solution for these eager, idealistic, young men was to build a financially profitable consulting practice that could finance their scientific interests.

The prospects for medical research thus looked very bleak in the United States around 1880, but the situation changed dramatically for the better within less than two decades. A few universities, medical schools, and research institutes received financial support, chiefly from private funds, to develop scientific programs modeled on the European examples. This profound change of attitude was probably in part an expression of America's coming of age, but it was accelerated by two independent kinds of influence that operated simultaneously at a critical time: the practical applications of the germ theory to the prevention and treatment of disease, and the emergence of social philanthropy as a result of the sudden accumulation of great wealth by a few American families.

Although scientific medicine had advanced on many fronts in Europe during the first three-quarters of the nineteenth century, it had not contributed much of practical use to either the prevention or the treatment of disease. Its main achievements had been in the description, classification, and natural history of pathological disorders. The new knowledge had made clinicians more competent in diagnosis and prognosis, but not much more effective in treatment. As a consequence of this limitation, scientific medicine was not really meaningful to the general public or even to the average run of practicing physicians. Many of these, in fact, considered that it had only a negative effect, because it discredited some of the time-honored practices by questioning their effectiveness and emphasizing their dangers. Such an attitude of therapeutic nihilism among scientific physicians was, of course, largely justified during most of the nineteenth century, but its unfortunate result was to retard the general acceptance of scientific medicine by destroying confidence in old practices without offering anything better as a substitute.

A profound change in public attitude occurred during the last quarter of the century, after the demonstration by Pasteur, Koch, and their followers that many types of disease were of microbial origin and could be prevented or treated by measures directed against the microbes. Lister's application of the germ theory to the control of surgical infections, the development of sanitary practices, and the use of vaccines for prevention and of immune sera for treatment proved that knowledge derived from laboratory research could be of practical usefulness. The germ theory thus provided the first convincing and obvious evidence that laboratory research was helpful, not only for understanding disease, but also, and more importantly, for the control of disease. Scientific medicine was widely accepted by the general public and by official bodies as soon as it became prescriptive instead of merely descriptive.

Despite the public and official recognition of its practical value, scientific medicine would probably not have developed as rapidly as it did if public funds had constituted its only source of support. The promotion of laboratory research would almost certainly have been handicapped by administrative difficulties and by the conservatism of medical and academic institutions. Fortunately, social changes that were then occurring in the United States provided sources of private funds to catalyze scientific programs and to experiment with new scientific institutions.

Great fortunes had been made during the second half of the nineteenth century, and a few men of wealth decided to devote very large sums to public ends during their own lifetimes. As they were eager that their philanthropies be rational and creative enterprises instead of mere charity, they sought the help of advisors committed to one or another type of social cause. Thus emerged one of the most striking social phenomena of our times—the use of private funds for the support and establishment of libraries, educational institutions, research laboratories, hospitals, and medical schools, as well as of concert halls, theaters, and artistic or charitable programs. The new philanthropists shifted the emphasis from traditional charity at the individual level to programs for social improvement.

Two cases of nineteenth-century social philanthropy had a direct relevance to the topics of this book: The Johns Hopkins institutions in Baltimore (Maryland), which prepared the ground for The Rockefeller Institute for Medical Research; and the Hoagland Laboratory in Brooklyn (New York), which provided Avery with experience in medical research.

At his death in 1873, the Baltimore Quaker merchant Johns Hopkins left about $7,000,000—then a very large sum—for the establishment of three institutions: a new type of university focused on research activities, rather than on didactic undergraduate teaching; a hospital with facilities for the investigation of disease; a medical school linked to both the university and the hospital, and therefore of academic character.[8] The university was established first; then an institute for research in pathology, even before the hospital was ready to receive patients; and, last, the medical school. William Henry Welch, who a few years before had despaired of ever finding facilities in America for medical research comparable to the ones he had known in Germany, was appointed the first director of the new Johns Hopkins Institute of Pathology in 1885, with the expectation that he would devote himself entirely to research and teaching in a university environment. Thus began the phenomenal career which made him the architect of American scientific medicine. Although his base remained The Johns Hopkins Medical School, he played a crucial role in the creation of The Rockefeller Institute for Medical Research and of many institutions of a national character. The Rockefeller Institute was organized by Simon Flexner, who had been one of Welch's favorite pupils; in addition, Welch served as chairman of the Board of Scientific Directors of the Institute from 1901 to 1933.

Whereas The Johns Hopkins University, Hospital, and Medical School have long been in the limelight, few persons know of the Hoagland Laboratory, which was incorporated in 1887. It has now been discontinued and is almost completely forgotten, except by those who know that Avery worked in it from 1907 to 1913. Yet, the Hoagland institution is historically important for having been the first privately endowed American laboratory focused on bacteriological research. It was dedicated in 1888, the same year as the Pasteur Institute in Paris.[9]

Cornelius Nevius Hoagland (1818–1898) was a physician who practiced medicine for 13 years. Then he went into business with his brother, and made a sizable fortune by promoting baking soda and creating the Royal Baking Powder Company. In 1884, his grandson died of diphtheria, and this tragedy reawakened his medical interests. He was aware of the spectacular advances made in medical bacteriology, and came to believe that this science was the only one that offered mankind a real hope against disease. He returned to medical practice and endowed a laboratory devoted to bacteriological research and teaching. Its first director was Dr. George M. Sternberg, major in the Medical Corps of the U.S. Army, who had achieved international fame for his work on yellow fever. He was appointed in 1888 and remained Hoagland's nominal head until 1893, when he resigned to become Surgeon General of the U.S. Army.

The Certificate of Incorporation of the Hoagland Laboratory stated that its objectives were "the promotion of medical science and the instruction of students in special branches thereof." William Henry Welch was to have been the main speaker at the dedication of the new buildings on December 15, 1888, but he was prevented by a prior engagement. His place was taken by Dr. H. Newell Martin, professor of biology at The Johns Hopkins University, an Englishman who was one of the first men in the United States to devote his entire time to teaching and research on a medical subject. One of Martin's themes in his address was that "whereas in

Europe science was often directed from a government office and central-ized under bureaucratic control, in this country the . . . endowment of laboratories was being attained in a far better way – by private generosity rather than by public subsidy." "Science," Martin said, "cannot for any long period advance safely in chains even if those chains be golden. Through private endowments – trusts as they are for the public welfare – American science promised to attain a variety and independence of thought such as no national science had ever attained in the past."[10] Martin's speech sounded a theme that was to be of great importance for the development of medical research in America, and that is still timely today: the independence of the scientist from bureaucratic control.

Because of shortage of funds, the Hoagland Laboratory was incorpo-rated into the Long Island College of Medicine, which eventually became the College of Medicine, State University of New York, Downstate Medi-cal Center. But one fact deserves to be restated before taking leave of this small, pioneering institution. In 1899, Dr. Sternberg, then U. S. Surgeon General, stated that, as far as he knew, "The Hoagland Laboratory is the first laboratory in the United States erected, equipped and endowed by private means for the sole purpose of bacteriological research."[11] Admit-tedly, bacteriological research and teaching had been conducted at the Carnegie Laboratory of Pathology in New York City as early as 1885, but that laboratory, as its name indicates, was built primarily for pathology; the bacteriology done there was largely incidental.

Avery, who worked in the Hoagland Laboratory for approximately six years, was fond of saying that the professional associations and the intellec-tual freedom he had enjoyed there had contributed greatly to his scientific development.

The Rockefeller Institute for Medical Research

The organizational patterns of medical research in the United States appeared fairly well established by the last decade of the nineteenth century. Research programs had been created and financed, first in a few privileged medical schools, hospitals, and universities, then throughout the land; public funding had followed private funding. The development of scientific medicine probably would have continued along the same course as part of an orthodox academic tradition, if it had not been for the unexpected impact of a layman. His name was Frederick Taylor Gates, a Baptist minister who acted as adviser to Mr. John D. Rockefeller in matters of philanthropy. In addition to his crucial role in the creation of the Institute, Gates engaged in many other activities that influenced the course of modern medicine, for example the establishment of the Rockefeller

Foundation and the development of "full-time" academic medicine.

The events, thoughts, and discussions that led to the establishment of the Institute have been described in detail elsewhere, but some of their aspects must be mentioned here, because they help to explain the social and scientific atmosphere from which emerged the Institute's and Avery's accomplishments.[12]

Gates had always been interested in medical subjects. He was a physician's son, and had observed during his ministry as pastor of a struggling Baptist church in Minneapolis that physicians were usually unable to deal with serious medical problems. In 1897, he resolved to familiarize himself with the state of medicine, and was advised by one of his young friends, who was a medical student, that the most readable and competent source of information was Osler's textbook *The Principles and Practice of Medicine*.

Osler was acknowledged everywhere at that time as a physician of immense learning and ability, and also as a great teacher and skillful writer. His thorough training in pathology enabled him to supplement his clinical descriptions with vivid pictures from first-hand knowledge of the anatomical lesions characteristic of each disease. Furthermore, he was extremely fond of medical history and of its heroes; he was also familiar with the allusions to medicine that occur in poetry, novels, and other forms of literature. Finally, he was so adept in the expression of his thoughts that, although his textbook was intended for medical students and physicians, it had a literary and human quality that made it palatable to lay readers, especially to Frederick Gates.

One aspect of Osler's book that was regarded as a weakness by many physicians was skepticism concerning the prevalent forms of therapy and the use of drugs, in particular. But it was this therapeutic nihilism that impressed Gates and motivated him to work for the establishment of a medical research institute and, later, for other massive investments in biomedical research. Gates himself has given a detailed account of his reaction to Osler's book in a memorandum that he prepared in 1897 from his own recollections and from Mr. Rockefeller's private files:

Osler's *Principles and Practice of Medicine* is one of the very few scientific books that I have ever read possessed of literary charm. There was a fascination about the style itself that led me on and having once started I found a hook in my nose that pulled me from page to page, and chapter to chapter, until the whole of about a thousand closely written pages brought me to the end. But there were other things besides its style that attracted and constantly, in fact, intensified my interest. I had been a sceptic before. . . . This book not only confirmed my scepticism, but its revelation absolutely astounded and appalled me. . . . I found, for illustration, that the best medical practice did not, and did not pretend to cure more than four or five diseases. . . . It was nature, and not the doctor, and in most instances nature practically unassisted, that performed the cures. . . . [Osler's] chapter on any particular disease would begin with a profound and learned discussion of the definition of the disease, of its extension throughout the world, of the history of discovery about it, of the revelations of innumerable postmortems, of the symptoms, cause and probable results of the disease, and the permanent complications and consequences likely to follow, but when he came to the vital point, namely, the *treatment* of the aforesaid disease, our author . . . would almost invariably disclose a mental attitude of doubt and scepticism . . . about all that medicine up to 1897 could do was to nurse the patients and alleviate in some degree the suffering. Beyond this, medicine as a science had not progressed. I found further that a large number of the most common diseases, especially of the young and middle aged, were simply infectious or contagious, were caused by infinitesimal germs. . . . I learned that of these germs, only a very few had been identified and isolated. . . .

When I laid down this book, I had begun to realize how woefully neglected in all civilized countries and perhaps most of all in this country, had been the scientific study of medicine . . . while other departments of science, astronomy, chemistry, physics, etc., had been endowed very generously in colleges and universities throughout the whole civilized world, medicine, owing to the peculiar commercial organization of medical colleges, had rarely, if ever, been anywhere endowed, and research and instruction alike had been left to shift for itself dependent altogether on such chance as the active practitioner might steal from his practice. It became clear to me that medicine could hardly hope to become a science until medicine should be endowed and qualified men could give themselves to uninterrupted study and investigation, on ample salary, entirely independent of practice. To this end, it seemed to me an Institute of medical research ought to be established in the United States. Here was an opportunity, to me the greatest, which the world could afford, for Mr. Rockefeller to become a pioneer. . . . I knew nothing of the cost of research; I did not realize its enormous difficulty; the only thing I saw was the overwhelming need and the infinite promise, worldwide, universal, eternal. Filled with these thoughts and enthusiasms . . . I dictated to Mr. Jones, my secretary, for Mr. Rockefeller's eye, a memorandum in which I aimed to show to him, the to me amazing discoveries that I had made of the actual condition of medicine in the United States and the world as disclosed by Osler's book. I enumerated the infectious diseases and pointed out how few of the germs had yet been discovered and how great the field of discovery, how few specifics had yet been found and how appalling was the unremedied suffering. I pointed to the Koch Institute in Berlin and at greater length to the Pasteur Institute in Paris. . . . I remember insisting in this or some subsequent memorandum, that even if the proposed institute should fail to discover anything, the mere fact that he, Mr. Rockefeller, had established such an institute of research, if he were to consent to do so, would result in other institutes of a similar kind, or at least other funds for research being established, until research in this country would be conducted on a great scale and that out of the multitudes of workers, we might be sure in the end of abundant rewards even though those rewards did not come directly from the Institute which he might found.[13]

From the beginning, Gates had visualized that medical research could best be carried out by "an institution in which whatever practice of medicine there is, shall be in itself an incident of investigation." Mr. Rockefeller was much impressed by this view, and promised to provide funds for such an institution. He was the more receptive to the idea because his first grandchild, John Rockefeller McCormick, died of scarlet fever on January 2, 1901, when three years old. Mr. Rockefeller was shocked to learn from the doctors that the cause of scarlet fever was unknown and that there was no method of treatment for the disease. It is of interest that C. N. Hoagland also had been motivated to found the Hoagland Laboratory by the death of his grandson from diphtheria, another acute bacterial disease.

The Rockefeller Institute for Medical Research was incorporated that same year. Its financial resources were small at first and were used to support laboratory investigators in medical schools and hospitals throughout the country. In 1903, the Institute established its own laboratories in temporary quarters on Lexington Avenue in New York City. In 1906, the laboratories were moved to the present site, as discussed earlier. The genesis, history, and administrative structure of the Institute have been described in several books,[14] and further details will be published in a biography of the first director, Simon Flexner, now in the course of preparation by Dr. Saul Benison. One aspect of this history of direct relevance here is the scientific and administrative philosophy that presided over the organization of the Institute, because it was at least as important as the generous funding and the talent of the scientists in determining the continued success and the unique character of the enterprise.

As a clergyman, and as the externalized conscience of the Rockefeller family, Frederick T. Gates saw medical research both as a search for preventive and therapeutic measures against disease and as a way to probe into man's nature. Addressing The Rockefeller Institute scientists on the tenth anniversary of the opening of the laboratories, he told them that their work went far beyond the study of disease. In his words, "Your vocation goes to the foundation of life itself."[15] On another occasion, he asserted that the Institute could be regarded as a "technological seminary" that would provide material for a new form of religion. "I am now talking of the religion, not of the past, but of the future. . . . As this medical research goes on you will . . . promulgate . . . new moral laws and new social laws, new definitions of what is right and wrong in our relations with each other. . . . You will teach nobler conceptions of our social relations and of the God who is over us all."[16] This euphoric and somewhat misty view of the social role of science was, of course, foreign to most of the Institute's scientists; yet it contributed to the success of the enterprise by dedicating it to the search for knowledge, rather than to immediate practical applications.

The broad view of medical research held by the planners of the Institute made it easier to avoid short-term utilitarianism. The Rockefeller family and Mr. Gates had the wisdom not to expect quick results and not even to "cherish extravagant dreams"; it is truly remarkable that, even though they had little, if any, training in science and few contacts with scientists, they rapidly came to realize that the best chance of success for the Institute was to approach medical research from a theoretical point of view, instead of looking for practical solutions to specific clinical problems. In the expressed view of Frederick Gates, the chief hope of the founders was that the Institute would serve as an example, and thus indirectly contribute to the making of discoveries elsewhere. In fact, this hope was rapidly fulfilled, because a multiplicity of other institutions patterned more or less after The Rockefeller Institute were soon created to deal with the various aspects of medical research.

A few remarks concerning the initial peculiarities of the Institute will suffice to illustrate how the conceptual breadth of its organization enabled it to evolve rapidly with the changes in science and thus to remain highly productive, even though the problems of infectious disease, which had motivated its creation in 1901, progressively became of less social importance.

Whereas the European institutes of medical research each were built around a single remarkable scientist, such as Pasteur, Koch, Ehrlich, or Pavlov, The Rockefeller Institute was organized as an association of talented division chiefs. The authority of the director, Simon Flexner, was very great indeed, but it was an authority of administrative, rather than scientific, nature. In Flexner's own words, the Institute "was not confined in its growth by the interests . . . of a commanding personality; it could look forward to a broader foundation of science . . .; its usefulness could not be so seriously impaired by the death or retirement of one man."[17] There is no doubt that such a structure, in which scientists were selected for what they had produced or for what interested them and what they might contribute, rather than because of the discipline they represented, made it easier for the Institute to make rapid adaptations to changes in the social and scientific atmosphere.

Even though the initial focus of The Rockefeller Institute was clearly "medical" research, the documents concerning its organization made few references to specific diseases. Laboratories were organized around investigators selected for their intellectual gifts and representing highly diversified areas of medical science. Microbiology and pathology were given a prominent place at first, because these sciences had produced the most spectacular results at the time the Institute was organized. However, physiology, experimental surgery, and chemistry were emphasized from the very beginning. As financial resources increased and as gifted investigators revealed new, promising fields of research, the range of specialities represented at the Institute came to include almost any kind of science that might have a bearing on health and disease — from the study of populations to that of physical laws.

The scientific basis of The Rockefeller Institute was thus initially, and has remained ever since, broader than that of the other institutions created before or after it to deal with medical research. This catholic approach is the more remarkable because the scientists who were responsible for the Institute's organization had been trained in specialized laboratories abroad, particularly in the then-famous German institutes of pathology, physiology, or chemistry. Despite this pervasive German influence, the pattern of organization adopted for The Rockefeller Institute was closer to the ideal of science that had been formulated in France by Claude Bernard. Of special relevance in this regard is Bernard's last book, *Leçons sur les phénomènes de la vie communs aux animaux et aux végétaux*,[18] because its very title conveys the belief that the fundamental processes of life are the same in all types of creatures, from microbes to man. This scientific philosophy was central to the organization and operations of The Rockefeller Institute. It led in 1910 to the appointment of the physiologist Jacques Loeb as head of a new department of experimental biology, created to deal with the contributions of physics and chemistry to all the fundamental processes of life in all kinds of organisms. The scope of the

Institute was eventually further enlarged by the addition of units devoted to animal and plant diseases. The Rockefeller Institute thus came to be involved in most aspects of biological research.

The Rockefeller Institute Hospital

There was nothing original in the idea that a hospital should be established in close association with the Institute laboratories. The originality of the enterprise emerged with the appointment of the first director and physician-in-chief, Rufus Cole, who insisted that the Hospital should be not an annex of the laboratories, but an independent unit completely equipped to conduct its own research programs. This concept was so new at the time that it justifies a few details concerning its origin, especially in view of the fact that it had a profound influence on the subsequent evolution of clinical research in North America.

In his initial 1902 plan for the organization of the Institute, Simon Flexner had urged the establishment of a hospital, were it only to make sure that problems of human disease not be forgotten by laboratory scientists. He did not have a clear notion of how it should be organized, and merely stated, "The hospital should be modern and fully equipped, but it need not be large. It should attempt to provide only for selected cases of disease."[19] This phrase "selected cases of disease" obviously meant that the work of the Hospital was to be focused on specialized clinical research, rather than on general medical care.

Physicians who are scientifically oriented now take it for granted that clinical research implies laboratory research, but this view was still foreign to the most illustrious representatives of scientific medicine less than a century ago. Shortly after returning from his second trip to Germany in 1884, William Osler had stated that "the wards are clinical laboratories utilized for the scientific study and treatment of disease."[20] He shared the opinion held by most physicians that biological and chemical laboratories were needed only as diagnostic tools. Originally trained as a naturalist, Osler carried this attitude into medicine. As a scientific physician, he was primarily interested in the natural history of disease, and he derived his knowledge from the careful observation of signs and symptoms in patients. Medical science meant to him not experimental science, but clinical information garnered at the bedside and interpreted in the light of data obtained by bacteriological and chemical techniques, medical statistics, and especially post-mortem examination of previous, similar cases.

Despite Osler's prestige, the attitude toward medical research changed during his own lifetime. L. F. Barker, who succeeded him in the chair of medicine at The Johns Hopkins Medical School in 1906, had first had a career in laboratory research. Immediately after his appointment, he established, adjacent to his wards, biological and chemical laboratories intended for research by his clinical staff in the mechanisms of disease, rather than for diagnostic tests. Rufus Cole, who was the first head of Barker's biological laboratory, took advantage of the situation to carry out experimental studies on typhoid fever.

From the beginning, the Rockefeller Hospital had been "designed wholly for research in clinical medicine. Laboratories were provided on the same scale as beds for the patients, and in proximity to them,"[21] but it had been assumed at first that the Hospital would be a place where the Institute pathologists, physiologists, and bacteriologists could test ideas developed in their own laboratories, or where they could obtain specimens for their own investigations. Under such a system, the junior hospital interns were expected to serve as medical attendants looking after patients for the sake of scientific investigators — under the guidance of the physician-in-chief and visiting physicians.

When Rufus Cole was invited to be physician-in-chief in 1908, he proposed an entirely different plan. He wanted the physicians responsible for the care of patients to be given the right to investigate the fundamental mechanisms of disease as deeply as their training allowed. Collaborative projects between the Institute laboratories and the Hospital might, of course, be extremely valuable but, according to Cole, the real point of a research hospital was that physicians could engage in fundamental studies on the problems they dealt with in the wards.

Cole's point of view eventually prevailed, but there was much resistance to it within the medical community, on the grounds that medicine was more an art than a science. Whether medicine is an art or a science *sui generis* was still hotly debated a few decades ago, and several of the senior members of The Rockefeller Institute Hospital staff contributed actively to the debate. In practice, however, from the time of its opening, the Hospital operated in exactly the same spirit as the other Institute departments, except that the investigations carried out in its laboratories were chiefly derived from clinical problems.[22] During the early years of the Hospital, the main problems under study were lobar pneumonia, syphilis, poliomyelitis, rheumatic fever, heart disease, renal diseases — all pathological conditions that were prevalent at the time. The patients never numbered more than 60, and were selected as typical of the diseases under study. In subsequent years, other pathological conditions were studied when they became of greater social or scientific interest.

Granted the inescapable necessity for clinical specialization, there prevailed among the Hospital staff a breadth of intellectual interests very similar to that found in the rest of the Institute. A large percentage of the staff did not have medical degrees, but instead were Ph.D.'s in theoretical sciences, such as chemistry, physiology, or microbiology. Even more remarkable was that the head of the department of renal diseases, Donald D. Van Slyke, was not a physician, but a chemist.

Dr. Thomas M. Rivers, who replaced Cole as the director of the Hospital in 1937, has this to say concerning the place of Ph.D.'s in clinical medicine.

Although Van Slyke was a Ph.D., he had charge of all the kidney cases in the hospital, and over the years I must say that he was a better physician as far as his handling of nephritis and nephrosis was concerned than most M.D.'s. Because Van Slyke was a Ph.D., he couldn't sign orders in the order book for medicine and drugs, nor could he order tests. However, most of the orders carried out by M.D.'s on the service were usually done at Van Slyke's suggestion. . . . As far as I am concerned no one in the United States has done as much as Donald Van Slyke to unravel the riddles regarding the physiology and diseases of the kidney.[23]

If I may be permitted to introduce a personal note, I, too, acted for a few years as head of a clinical department of tuberculosis, even though I do not have an earned medical degree.

While there was no real rivalry between M.D.'s and Ph.D.'s in the Hospital, there were frequent discussions concerning the comparative value of the two forms of training as a preparation for medical research. Here, again, I shall let Dr. Rivers, who was an M.D., state the problem in his own words:

The bald fact is that the Ph.D.'s never felt that the M.D.'s were scientists. Just recently, I heard a young doc at the hospital complain to one of the old Ph.D.'s at lunch that all the M.D.'s went around trying to pass themselves off as Ph.D.'s and he wondered if it had always been so at the Institute. "No," replied the old Ph.D. "Thirty years ago I tried to pass myself off as an M.D. Then they were the kingpins around here, you know."

In fact, Dr. Rivers continued,

. . . a large number of the doctors in the hospital . . . were made members of the National Academy of Sciences. Cole was one, Avery was one, Dochez was one, and I was one, and there were others. I would like to make it clear that we weren't elected because we were M.D.'s . . . ; we were elected on the basis of our proficiency in one of the basic sciences.[24]

As it turned out, the physicians who had responsibility for patients on the wards worked not only on problems directly related to the disease which was their particular clinical concern, but also on problems of broader scientific scope. For example, it was while working on the causative agent of lobar pneumonia — the pneumococcus — that Avery's group discovered the role of DNA in the transfer of genetic characteristics. In addition to Avery himself, four of the members of the department who made fundamental laboratory contributions to this problem (Dawson, Alloway, Mc-Leod, and McCarty, see Chapter Eleven) were young physicians who had exacting ward responsibilities.

Thus, The Rockefeller Institute and its Hospital symbolize the explosive evolution — or, more exactly, the revolution — which began to transform American medicine at the turn of the century. Two different kinds of changes occurred during the first two decades of the twentieth century. The so-called Flexner Report (prepared by Abraham Flexner, brother of Simon Flexner) brought about an improvement in medical education.[25] The Rockefeller Hospital was one of the institutions that served as a model for the systematic use of the experimental method in the study of clinical problems. By 1920, the American medical establishment was committed to high standards of education and to the development of research. Not only had medicine become more scientific; its practitioners were discovering new general laws of biology, and even contributing to the advancement of other sciences.

Avery's career provides a spectacular example of this medical evolution, ending in a scientific revolution. Immediately after completing his medical training, he entered into private practice and used the empirical healing arts that constituted the medicine of his school years. He elected to abandon medical practice for laboratory work, and dedicated himself to scientific studies bearing directly on the understanding and control of disease. Finally, he contributed theoretical knowledge that revolutionized certain biological concepts and that may eventually affect the practice of clinical medicine.

From Research Institute to University

Because the founders of The Rockefeller Institute aspired to approach medical research on a very broad front, they had to provide staff and resources for a large diversity of scientific disciplines. Furthermore, they adopted a pattern of organization that gave almost complete autonomy to each of the scientific departments. There was a danger that such a combination of diversity and autonomy would produce a heterogeneous institution, but The Rockefeller Institute was, in fact, a remarkably well-integrated scientific organism.

The integration of the Institute's various departments was facilitated at first by the relatively small size of the initial staff, and also by the sense of community that resulted from scientific pioneering—against the doubts, and even the hostility, of a large part of the medical establishment. Of greater and more lasting importance for the intellectual unity of the Institute, however, was the administrative wisdom of its first director, Simon Flexner. While acting as Eastman Visiting Professor at Oxford University in 1937, Flexner discussed the general problem of organization of university clinics in words that obviously reflected his long experience with the problems of interdepartmental collaboration at the Institute. He emphasized that the design and arrangement of buildings should encourage "a spirit of easy and free cooperation. . . . At The Rockefeller Institute, covered corridors, heated in winter, connect hospital and general laboratories and passage from one to the other is made so easy that the effect is of one common building for all."[26] He could have mentioned also the conferences held every Friday afternoon, during which one of the staff members presented the results of his investigations and at which all members of the staff were expected to be present. Flexner himself was very much in evidence at these staff conferences, always sitting in the front row.

Probably most important of all, there was the lunch room with its comfortable chairs, its baguettes of French bread, its fresh butter, and its endless supply of coffee served by quiet and obliging waitresses. The tables accommodated eight persons, the right number to generate a conversational social unit in which all could participate. More often than not, the conversation at one table was dominated by one particular person. If Thomas M. Rivers was present, the talk at his table was likely to be about viruses or hospital management; with Alfred E. Cohn, it was about the history or philosophy of science; Avery spoke little, but listened and asked a few pointed questions bearing on some scientific problem that preoccupied him. It was in the lunch room that I was first introduced to Avery and that we engaged in the conversation which eventually brought me to the Institute.

Even Paul de Kruif, who did not think much of the Institute, wrote with warmth of its lunch room:

After a drab morning . . . at the lunch break there was balm for my discouragement. Here I could listen to the scintillating talk of my betters, a scientific elite, a bevy of bacteriological, biological big names. I was thrilled to sit at Jacques Loeb's table, listening to that parent of fatherless sea urchins. . . . Those luncheon sessions were a kindergarten in my stumbling study of character. I never tired of listening to the philosophy of Alex Carrel. . . . Then at the luncheon table there might be Dr. Peyton Rous, refined, gentle, exquisitely cultured. . . . In this refectory there was an air of solemnity to be expected and appropriate to the unveiling of mysteries.[27]

Wherever they meet, the alumni of The Rockefeller Institute evoke with gratitude and fondness the lunch-room conversations which familiarized them with the skills of their colleagues. This experience not only provided factual information and a broader perspective of science; it also generated collaborative projects with fellow scientists in other disciplines. It can be said without exaggeration that there never was a symposium — in the etymological sense of the word, namely, a convivial meeting for drinking — that was more scientifically productive and intellectually pleasurable than those held daily in the lunch-room of The Rockefeller Institute, though coffee and ideas were the only intoxicants.

The Hospital had its own social institutions that facilitated scientific contacts. Tea was served at 4:30 every afternoon in the residents' living room, and this was an occasion for much professional interchange. There was also the Hospital journal club that met for dinner, with wine or beer, every other Monday from October to May. It was started by Alfred E. Cohn, who faithfully presided over all its meetings until his retirement in 1944, and who introduced a subtle formality with certain traditions — for example a special menu of English (Dover) sole for the first meeting in October; of shad, asparagus, and strawberries in the Spring.

At each meeting, three or four papers were presented under a tacit set of rules. Except in very special cases, one was not supposed to review papers published in journals that we were all expected to have seen, such as the *Journal of Experimental Medicine,* the *Journal of Biochemistry,* the *American Journal of Physiology,* or the *Journal of Clinical Investigation.* The papers selected had to have some bearing on experimental medicine very broadly conceived, preferably from the outer margin of the speaker's research activities. Discussions were intense, and every member of the club could participate in them because the group was rather small — fewer than 20 during the period of which I speak. It was bad taste, however, to use the occasion for speaking of one's own research work and even for presenting results obtained in the Hospital departments. Finally, Dr. Cohn made it a point *not* to forewarn the persons who would be called on to speak — a practice that caused much anguish among junior scientists. Month after month, one came to the journal club dinner, wondering whom Dr. Cohn had in mind for that particular evening. One was expected to have a new paper to report at any meeting and, furthermore, one had to be prepared to present it with style, because all the heads of departments, as well as the Hospital director — first Dr. Cole, then Dr. Rivers — attended the meetings, and certainly used the occasion to evaluate their junior staff.

Many examples could be given to illustrate the community of interest that emerged from the focusing of thought on medical research and from the ease of intellectual contacts within the Institute. Immediately after the reference to the design of buildings that Flexner made in his Oxford speech, mentioned above, he went on to say, "Science thrives best . . . where research is in the air."[28] But, he pointed out, each place has a special research atmosphere of its own, determined by the goals of the institution, and he chose the case of Donald D. Van Slyke to illustrate how the "atmosphere of research" influences scientific creativity:

> Van Slyke's original training was in organic chemistry, but he early showed a special talent for physical chemistry. Had he developed outside the hospital, his interests would have been directed to the application of physicochemical methods to structural chemistry. In the clinical laboratory he was confronted with the problem of acidosis in diabetes and he concentrated his attention on the development of methods of blood analysis, which resulted in his discoveries in acid-base regulation in health and in disease. As time went on other contributions to quantitative clinical chemistry came from his laboratory, the responses to the conditions arising within the clinic and the medical atmosphere surrounding him.[29]

Flexner further illustrated the importance of the research atmosphere by referring to the programs on animal and plant diseases at the branch of the Institute that had been created in Princeton in 1914. These programs were conducted in two separate divisions, organized in such a manner that each was complete and independent of the other. The staff of each worked in the atmosphere created by the subject, a clinical atmosphere of sick animals or sick plants. On the other hand, there was close cooperation between the two divisions, because they adjoined each other.

Human diseases, animal diseases, and plant diseases are but a few among the research programs of the Institute that could have been selected to illustrate how its administrative structure made it possible to reconcile a unity of intellectual atmosphere with great diversity of scientific disciplines and complete autonomy of departments. It would probably have been difficult to maintain this unity if the work of the Institute had been directed to applications of an immediately practical nature, but this was not the case. Many new products, techniques, and gadgets of practical utility were, of course, developed by the Institute scientists. However, even in the programs directly focused on practical problems, for example on a particular human disease, the emphasis was primarily on scientific understanding — with the expectation that this understanding would eventually pave the way for practical applications, as it did in many cases. The motto of the Institute is *Pro bono humani generis;* its fundamental philosophy has always been that the most important contribution that science can make to human welfare is the kind of knowledge that facilitates a more intelligent conduct of life.

The Institute was not listed as an educational institution during the first half-century of its existence, but many young men, usually in their twenties, joined it to work under the guidance of its scientists, either on fellowships or as junior members of the staff. In any research program, teaching inevitably becomes a desirable, as well as a pleasurable, aspect of the relationship between leader and neophyte. Thus, although the Institute did not give academic degrees, it helped many young men and women to achieve superior preparation for a career in the field of science they themselves had selected. More importantly, perhaps, it provided an atmosphere in which they could discover themselves by being exposed to the wide range of scientific disciplines and intellectual attitudes represented in the laboratories, during staff conferences, and in the lunchroom every day.

From its very beginning and throughout its existence, The Rockefeller Institute thus had certain general characteristics of a university. Its diversity of disciplines and of administrative structures have probably played a large part in assuring its continued scientific productivity and in making it adaptable to changes in science. To summarize:

Instead of being focused on practical problems, the Institute cultivated a broad approach to the understanding of biological principles, using whatever concepts and techniques were available at a given time. Instead of being built around a single remarkable but dominant personality, as were several of the other medical research institutes, it was organized as a commonwealth of scholars representing a great diversity of scientific disciplines. Instead of training students by conventional teaching methods, it gave young scientists the opportunity to discover their own tastes and talents by working in association with a self-selected master. Thus, while functioning as a research center focused on medical problems, the Institute displayed some of the most desirable attributes of a university. Few difficulties of adaptation were experienced when, in 1955, it was transformed into The Rockefeller University and had to enlarge still further the range of its research and teaching activities.

CHAPTER THREE

CHEMISTRY IN
MEDICAL RESEARCH

Chemistry at the Birth of The Rockefeller Institute

The Rockefeller Institute for Medical Research was conceived at a time when infectious diseases were the most important medical problems in all countries of Western civilization. Microbiological sciences were then the most glamorous field of medical research because of their spectacular contributions to the understanding and control of pathological processes. This period has been called the golden era of microbiology, because each year saw the discovery of new infectious agents and of new preventive and therapeutic methods.

When Frederick T. Gates read Osler's textbook in 1897, he had been much impressed by the fact that "a large number of the most common diseases, especially in the young and middle-aged," were caused by microbial agents. In the plan he submitted to John D. Rockefeller for the promotion of medical research in the United States, he used as models the Pasteur Institute in Paris and the Koch Institute in Berlin, because these two institutions were primarily concerned with infectious diseases.[1] It would have been natural, therefore, to focus the resources of the new Institute on microbiology and immunology, as had been done in Paris and Berlin, and shortly after at the Lister Institute in London and the Kitasato Institute in Tokyo.

As soon as The Rockefeller Institute was created, the post of director was offered to Theobald Smith, who had achieved fame in infectious pathology by his brilliant studies on Texas fever of cattle. Smith declined the offer for personal reasons, but as he was then a member of the Board of Scientific Directors of the Institute, he took this opportunity to express the opinion that the best policy for the Institute was to concentrate on "the study of infectious diseases from all points of view. They are the great threatening dangers of our present social system."[2] Following Smith's refusal of the directorship, another member of the Board of Scientific Directors, T. M. Prudden, drew up a statement of policy for the Institute, and he, too, came to the conclusion that the major aim should be, at first, the investigation of infectious diseases.[3]

In 1902, Simon Flexner was appointed director. As has been mentioned, he had been one of W. H. Welch's closest associates, and had made his scientific reputation as a pathologist and bacteriologist. Yet, even though he had been concerned almost exclusively with experimentation and teaching in the field of infectious diseases, he decided that the scientific scope of the Institute should be broader than had been recommended by Smith and Prudden. From the time he assumed the directorship, he acted on the conviction that medical research would, in the future, become increasingly dependent on chemical knowledge. Evoking the early years of the Institute, he wrote later, "In those days of rapidly advancing immunology and chemotherapy, Ehrlich's Institute in Frankfurt was a great attraction. . . ."[4] Biology as a whole . . . was fast taking on a chemical guise. . . . The Rockefeller Institute took part in this growth by providing in its original organization for biological chemistry . . . in its organic and physical forms. Biophysics, the corresponding new discipline, was added later."[5] It is probable, however, that Flexner's belief that chemistry would play a crucial role in medical research went further back in time than Ehrlich's influence. It can be traced to his early association — first as a graduate student, then as a colleague — with Welch at The Johns Hopkins Hospital and Medical School.

Welch had begun to prepare himself for medical school in 1870 by serving as an apprentice to his father, who was a medical practitioner, but he did not enjoy the experience. The following year, he went back to Yale University, this time to study chemistry at the newly-created Sheffield School. He did well in his chemical studies and, according to his biographers, Simon and James Flexner, there is evidence that he had even then "an intuition of the role that science was to play in the healing art."[6] After completing his chemical studies at the Sheffield School, Welch studied medicine at the College of Physicians and Surgeons in New York City. There again he did well, and a letter of the time to his father reveals how intensely he responded to topics of a quasi-chemical nature that had no direct relevance to the practice of medicine. He remembered to the end of his life a lecture in which the professor of *materia medica* and therapeutics, Edward Curtis, stated that protoplasm was "the physical basis of life in all its manifestations animal or vegetable. . . . The theory was beautiful."[7]

Christian Herter is another physician who probably sensitized Simon Flexner to the importance of chemistry in medical research. Herter was a friend of Mr. Rockefeller and was himself a man of wealth. For sheer intellectual satisfaction, he had established and funded a chemical research laboratory adjacent to his medical office in his midtown Manhattan house. Since he was among the first persons to formulate plans for the proposed Institute, and served from the beginning on its Board of Scientific Directors, one can assume that he was influential in making chemistry an important part of the initial research program.[8]

Flexner was, in any case, so convinced that chemistry would play a central role in medical research that, before undertaking his new duties as director, he spent a year abroad to familiarize himself with "the rapidly advancing science of physiological chemistry."[9] He studied in Berlin "with Salkowski and, more important, with Emil Fisher, who was doing his basic work in the chemistry of animal tissues and organs."[10] Upon his return to New York, his first administrative move was to appoint P. A. Levene as head of the chemistry laboratory. Levene was a Russian who had worked in Emil Fisher's laboratory and who "brought something of the problems and atmosphere of that exciting place to the infant institute."[11]

After Rufus Cole was appointed director of the Hospital in 1909, he also decided that he needed chemical knowledge to carry out his new duties as a leader of medical research. He was primarily a clinician, with some laboratory experience in pathology and bacteriology, but, instead of taking time out to strengthen his experience in these fields, he spent the year 1909–1910 working in Levene's department of chemistry at the Institute during the period when the Hospital was being built. It was at that time that he became acquainted with the organic chemist Donald D. Van Slyke, whom he brought to the Hospital a few years later.

In 1927, John D. Rockefeller, Jr. presented to the Institute a magnificent painting of the chemist Antoine Lavoisier, by Jacques Louis David. The painting now occupies the place of honor in the University library, located in the Welch building. There could not be a truer symbol of the Institute's scientific vocation, and of William H. Welch's influence on its destiny. It was Lavoisier who placed chemistry at the center of the contemporary biological sciences, and it was Welch who, indirectly through Simon Flexner, committed The Rockefeller Institute to the chemical view of medical research.

Chemistry as a Research Tool

A possible title for this chapter might have been "Better Life Through Chemistry," to emphasize the importance of the chemical approach in the development of scientific medicine. Unfortunately, the phrase has acquired a meaning far too narrow for what I have in mind. When it was first introduced as a publicity slogan by a large chemical firm a few years ago, it referred to the social changes that had resulted from the mass production, at fairly low cost, of a wide range of synthetic products such as plastic gadgets, artificial fabrics, food additives, fertilizers, and pesticides. Then the phrase was adopted by the youth culture to denote the various kinds of emancipation that could be achieved with contraceptives and with the immense variety of stimulating, relaxing, or mind-expanding substances produced in the laboratory. These examples provide obvious illustrations of the way in which modern life has been affected by chemical innovations. But the transformations of medicine by chemistry have been even more profound, and many of them have taken place in indirect ways that escape attention.

A spectacular demonstration of the beneficial role that chemistry has played in medicine is provided by the improvements it has made possible in the design, production, and use of medicinal drugs. The old saying that physicians put drugs of which they know little into human bodies of which they know nothing is no longer quite as true as it used to be. Many drugs are now specifically designed by chemical synthesis to fit certain physiological needs or to produce desired reactions; furthermore, methods are available to anticipate the biological and psychological effects of drugs even before they are released for general use. Also thanks to chemistry, vitamins and hormones have been isolated, identified, and synthesized; as a result, they can be used to regulate vital functions according to nature's own ways. Chemistry has put at the disposal of the physician powerful substances that enable him to exercise much control over physical and mental processes, in health and in disease.

The link between medicine and chemistry has been further strengthened by the chemical study of substances and systems derived from living organisms, and by the physiological study of the responses that the living organisms themselves make to natural and synthetic compounds that have biological activities. Chemists and biologists use a common scientific language when they study how a class of chemical substances, derived from a given biological system, sets in motion a certain type of physiological response.

One of the most interesting aspects of the chemical approach to medical research has been the possibility of reaching into all the regulatory processes that keep living organisms in a functioning order. The importance of such chemical regulation was clearly implied in Claude Bernard's concept that the stability of the internal environment is an essential condition of free life. As is now well understood, this approximate stability of the *milieu intérieur* is achieved through a complex system of chemical feedbacks, each step of which is controlled by hormones of exquisitely defined chemical constitution, functioning under precisely defined chemical conditions.

The phenomenal specificity of most biological processes is a special aspect of chemical regulation that deserves emphasis here because it played a dominant role in Avery's scientific achievements. The belief that biological specificity depends upon the molecular architecture of body constituents was accepted as an article of faith long before it could be scientifically demonstrated. For example, Paul Ehrlich had no precise knowledge of the mechanisms involved in the operations of antiseptics or antibodies when he boldly stated the problem in the famous phrases: "Antitoxins and antibacterial substances are, so to speak, charmed bullets which strike only those objects for whose destruction they have been produced."[12] And "only such substances can be anchored at any particular part of the organism which fit into the molecule of the recipient combination as a piece of mosaic fits into a certain pattern."[13]

A similar attitude found expression in the picturesque image used by Emil Fisher to account for the specificity of enzymatic reactions. According to him, the specific activity of each particular enzyme for a particular substance reflects a reciprocal fitness of structure comparable to the lock-and-key relationship.

Awareness of the chemical mechanisms of biological regulation and specificity was certainly an important factor in the formulation of medical research during the early days of The Rockefeller Institute. Paradoxically, however, chemistry came to dominate the intellectual atmosphere of the Institute, not through the achievements of the professional chemists, important as those achievements were, nor even through the preoccupations of the physicians concerned with the chemical aspects of physiological processes and of hormone, enzyme, or drug action, but through the vigorous personality of Jacques Loeb—a general biologist intent on promoting a philosophical theory of life based on physicochemical determinism.

The Chemical View of Life

Jacques Loeb was born in the Rhineland in 1859 and grew up during a period marked by the expression, among many young European scientists, of unqualified materialism. In 1845, for example, a quadrumvirate of rising German physiologists—H. Helmholtz, K. Ludwig, E. Dubois-Reymond, and E. W. Brucke—had committed themselves in a famous mutual oath to the demonstration that all bodily processes can be completely accounted for in physicochemical terms.[14] Although they tempered their materialistic view of life in their adult years, Loeb remained true to its most extreme form, and eventually became its outspoken missionary.

As a youth, Loeb had read extensively in the eighteenth-century philosophical literature on free will and consciousness. He entered the university in Berlin with the intention of becoming a philosopher. Soon, however, he realized that professors of philosophy could not answer the questions they delighted in posing, and he switched to science in the hope that he could solve, by observation and laboratory experimentation, the philosophical problems of the mind.[15] His life-long obsession with the metaphysical issue of free will can be traced to his youthful interest in Schopenhauer, with whom he shared a dogmatic conviction that individual freedom is illusory. Like Schopenhauer, also, he had a passionate desire to convey this unmitigated truth to others, an attitude that found expression in his assertiveness as a teacher and publicist.[16]

In 1880, at the age of 21, Loeb enrolled at the University of Strasbourg, from which he obtained an M.D. degree in 1884. There, he worked in a laboratory concerned with the localization of brain function, but again was disappointed; in his quest for the understanding of free will, he did not get from the neurologists any more enlightenment than he had from the philosophers. He then moved to the University of Würzburg, and joined a group of experimenters who were working on the borderline between biology and physics on the subject of tropism, or involuntary movement, in organisms. Among these men, he found at last a congenial spiritual home, because tropism was a kind of behavior that could be investigated in the laboratory by experimental techniques.

In Würzburg, he became acquainted with the young Swedish chemist Svante Arrhenius, or at least with his theory of electrolytic dissociation.[17] The discovery that complex chemical processes can be explained by simple physical laws converted Loeb to the view that physical chemistry would eventually explain all biological phenomena, including those of embryological development. While working from this point of view at the Naples biological station, Loeb succeeded in causing segmentation of sea-urchin eggs by altering the osmotic pressure of the fluid in which the eggs were immersed. This achievement was widely acclaimed as a triumph of mechanistic physiology, and made him known throughout the scientific world. His further studies on artificial parthenogenesis reinforced his physicochemical view of biological determinism and, in addition, increased his fame, especially among American biologists.

He moved to the United States in 1891, first to Bryn Mawr College, then to the University of California and the University of Chicago before joining The Rockefeller Institute in 1910. Unceasingly, he channeled all his knowledge and energy into the applications of physical chemistry to biological processes, and he became the most effective spokesman — indeed an evangelist — for the "Mechanistic Conception of Life." His book of that title was published in 1912, and is one of the landmarks of twentieth-century biology, if not by its purely scientific content, at least by the influence it exerted on several generations of biologists all over the world.

Loeb believed that the physicochemical view of life was the key to general biology and to scientific medicine, as well. After being invited to organize a new department at The Rockefeller Institute, he wrote Simon Flexner that he wanted to develop experimental biology "on a physico-chemical instead of on a purely zoological basis," and that "The experimental biology of the cell . . . will have to form the basis not only of Physiology but also of General Pathology and Therapeutics."[18] He affirmed this scientific philosophy with such vigor that he soon became one of the most influential members of the Institute staff. He left no doubt in the minds of those who listened to him — and one could hardly escape listening to him — that the only worthwhile kind of medical research was the investigation of simple biological systems by the methods of physics and chemistry.

The very logic of his physicochemical view of life led him from the study of living organisms to that of chemical constituents separated from biological materials, and eventually to that of simple colloidal systems — the simpler the better. At the time of his death in 1924, he was investigating the effects of various kinds of salts on gelatin under different conditions of acidity and alkalinity. He had chosen to work with gelatin not because he had a special interest in the biological role of that protein, but merely because it provided him with a colloidal system of simple composition. He believed, indeed, that he should have begun his study of life by investigating simple phenomena and substances, even though these were of limited biological importance, because it is "more logical to commence with the simple systems found in colloids than with such conditions as exist in protoplasm."[19]

Loeb's analysis of biological processes was thus an almost infinite regression. Believing as he did that one could not understand anything of psychological or physiological behavior unless one knew everything about molecular behavior, he tended to be contemptuous of orthodox biological and medical research. His dogmatic attitude on this score naturally irritated many of his Institute colleagues, but his intellectual prestige was so great that he had devoted admirers, especially among the young members of the staff, some of whom tried to ape his tart dialectic. In the words of Paul de Kruif:

> He gave the Institute a high scientific tone. . . . He was the peerless leader of the militant godless. . . . He was the exponent of scientific method as against the prevailing twaddle — that was his word — of medical science.
>
> "Medical science? . . . Dat iss a contradiction in terms. Dere iss no such thing. You should begin with the chemistry of proteins, as I do" he admonished his table mates in the Institute lunch room.[20]

Despite, or perhaps because of, his scorn for medical research, he had admirers even among physicians on the clinical staff of the Hospital. Many of them, however, "were filled with a kind of consternation"[21] on being told by him that they could not find anything useful about disease until they went much deeper into the intimate chemical mechanisms of the body. Alfred E. Cohn, who was then in charge of the cardiology department at the Hospital, reported their feelings later in his book *No Retreat from Reason*: "We . . . trained as physicians, were made unhappy. Loeb, the most accomplished, the most intelligent and we thought, the wisest man with whom it was our privilege to come in contact, as we did daily in our lunch room, we thought was laughing at us." The physicians on the clinical staff were willing to accept that "ultimately a human body is a mass of electrons," but they could not see how such knowledge would bring them "a step nearer to being able to do anything about pneumonia or cardiac disease."[22] On the whole, however, Loeb's faith in the physicochemical approach to biological problems found a favorable response throughout the Institute. Flexner, in particular, seems to have been receptive to that scientific philosophy.

The word "philosophy" is the proper expression to denote the nature of Jacques Loeb's influence on The Rockefeller Institute, because what he did and what he taught were determined more by his *a priori* philosophical view of life than by the study of actual processes in living things. Although he thought that he had abandoned metaphysics once and for all during his student years, in reality he had returned to it by his passionate espousal of mechanistic biology. He saw in physics and chemistry the only rational approach to the understanding of biological phenomena and of consciousness and free will. Without any possibility of scientific proof, he did not hesitate to affirm: "Not only is the mechanistic concept of life compatible with ethics; it seems the only conception of life which can lead to an understanding of the source of ethics."[23] Parenthetically, it is entertaining to compare this dogmatic statement with the views expressed by Frederick T. Gates when he told the scientific staff of the Institute (see Chapter Two) that medical research could be regarded as a new form of religion from which would emerge "new moral laws and new social laws"[24] and even nobler conceptions of God!

As mentioned earlier, Loeb's propensity to pronounce *ex cathedra* on the most fundamental issues of philosophy, morality, politics, and science irritated some of his colleagues, but not enough to nullify his influence on the conduct of biological research at the Institute and in other scientific institutions. By brazenly parading his mechanistic animus, he did more than anyone else to foster the belief that the most effective approach to biological and medical research is through physics and chemistry — a belief that has left an indelible stamp on the scientific approach to medical research at The Rockefeller Institute.

The year after Loeb's death in February, 1924, W. J. Osterhout was appointed head of the Institute's department of general physiology. Whereas Loeb had begun his scientific life as a zoologist, Osterhout was a botanist, but this difference was inconsequential, because the two men had the same fundamental interest — the desire to study the influence of physicochemical factors on the biological activities of isolated tissues or cells.

Like Loeb, Osterhout was fond of simple experimental models, which, irrespective of the extent to which they revealed new facts about life, provided a framework for thinking about biological problems. For example, he used large plant cells, such as those of the *Nitella* or *Valonia* genera, for the simple reason that they permitted direct observation of the passage of salts or dyes across their membranes under different experimental conditions. He also devised artificial cell models, which were somewhat similar to living cells in their ability to accumulate ions. The knowledge derived from the study of such simple systems — natural or artificial — guided physiological thinking about complex animal systems, such as those involved in renal activity, muscle contraction, or the transmission of nerve impulses. The subsequent scientific history of The Rockefeller Institute was further influenced profoundly by the fact that these simple systems lent themselves to analysis by physicochemical and mathematical methods, thus making it easier for professional physicists and chemists to become involved in biological problems.

Loeb and Osterhout naturally exerted a direct influence by their scientific discoveries, but more important in the long run was their indirect influence on the structure of the Institute's scientific staff. The very nature of their studies made both of them dependent on the collaboration of chemists and physicists, most of whom became interested in biological and medical research while applying their professional skills to experimental models. The physicists and chemists who came to work with Loeb and Osterhout remained on the Institute staff after the two masters had disappeared, and several of them became heads of new departments. As a result, the initial tendency of the Institute to approach medical problems through physicochemical methods was greatly reinforced. Whereas pathologists, bacteriologists, and virologists constituted the largest percentage of the staff during the early decades of the Institute, medical scientists progressively came to be outnumbered by chemists, physiologists, and biophysicists. In the 1940s, for example, there were six different laboratory groups working on the various ramifications of protein chemistry. Jacques Loeb would probably have taken great pleasure in learning that, in the 1960s, several biological departments that made intensive use of physicochemical methods elected to be listed under the broad heading of general physiology — the science that he had done so much to promote.[25]

Pure chemistry occupies only a rather small place in the present structure of The Rockefeller University, but the chemical approach is more dominant than ever in fields such as cellular biology, genetics, immunology, and experimental pathology. Chemistry as such has been replaced by molecular biology.

Interdisciplinary Thinking

The mere enumeration of the biologists, chemists, and physicists on the staff of the Institute does not give a true qualitative picture of its intellectual composition. More interesting is that, whereas representatives of each individual scientific discipline naturally retained their professional identification with regard to theoretical knowledge and laboratory techniques, many of them also developed lateral interests while collaborating on biological problems unrelated to the traditional preoccupations of their specialties.

For example, physical chemists working on electrophoresis became interested in the peculiar problems posed by the fractionation of blood-serum proteins; organic chemists working with nucleic acids learned to use biological systems to determine the role of those substances in the transfer and expression of hereditary characteristics. Conversely, bacteriologists acquired knowledge of molecular structure in order to account for immunological specificity; virologists learned to use ultraviolet absorption spectra of proteins and nucleic acids in their attempts to purify viruses.

The few examples just mentioned will suffice to illustrate how the diversity of problems and techniques generated in the Institute a spectrum of scientific interests much more complex and subtle than that defined by the traditional scientific disciplines. Superficially, these examples seem to be illustrations of what is now called interdisciplinary approach, but the actual scientific atmosphere did not result merely from the bringing together of specialists in different disciplines for the prosecution of well-defined projects.

On the one hand, most cooperative research within the Institute emerged spontaneously without administrative planning. It was commonly the outcome of lunchroom conversations—when clinicians and physical chemists discussed the possibilities of serum fractionation; when immunologists heard about Pauling's views of antibody protein folding; when I, who had been trained in soil bacteriology, learned of the need in certain clinical problems for specific chemical tests to which I could contribute by producing enzymes from bacteria. Thus, while most of the Institute's scientific projects were indeed interdisciplinary, few of them were the outcome of organization through administrative planning.

Even more important, however, was that some of the most striking examples of interdisciplinary interplay took place within the mind of each individual scientist, rather than among different scientists. The sensitization brought about by continued laboratory contacts made biologists think in chemical terms, and encouraged chemists to focus their thoughts and their techniques on the peculiarities of biological and medical problems. For example, it was because Avery and his medical associates had learned to think chemically that they were able to demonstrate the nucleic acid nature of the genetic material in pneumococci (see Chapter Eleven).

The mechanistic conception of life had led Jacques Loeb to formulate all biological problems in physicochemical terms, and this attitude enabled him to make a few startling prophecies. He believed, for example, in the possibility of producing mutations by physicochemical means; this was achieved in 1926 by the geneticist Hermann J. Muller through the irradiation of fruit flies with X-rays. In 1911, Loeb stated that the main task for students of heredity was to determine "the chemical substances in the chromosomes which are responsible for the transmission of a quality."[26] This was achieved three decades later by Avery and his group.

Loeb knew, of course, that the isolation and identification of the active substance in the chromosomes would require the use of sophisticated physicochemical methods; but he probably assumed that the work would be done by scientists trained in the physicochemical aspects of general physiology, the discipline he regarded as the fundamental biological science. He would therefore have been surprised to learn that the achievement had been the feat not of physicists, chemists, or general physiologists, but of physicians working in a hospital department dedicated to the study of lobar pneumonia.

In a way, the DNA story can be regarded as a vindication of Jacques Loeb's evangelism. The forcefulness with which he had preached the mechanistic gospel of life — to the point of intolerance — had created at The Rockefeller Institute an intellectual environment in which biologists and physicians took for granted that all their problems *should* and *could* be formulated in physicochemical terms and investigated by physicochemical methods. When Thomas Rivers stated that many of the Institute's M.D.'s were scientifically as competent as the Ph.D.'s and that several of them wanted to pass as such, he unconsciously reflected Loeb's view that there was no worthwhile medical science other than laboratory science. In fact, Loeb's intellectual influence has been so deep and lasting that it still conditions the attitudes of biologists and physicians who never saw him and are barely aware of his name. Because of it, ways of thinking about life, including human life, that do not involve a physicochemical approach have never found a congenial home within the walls of The Rockefeller Institute.

Exactly three decades after the publication of Loeb's *The Mechanistic Conception of Life,* genetics emerged as a physicochemical science from the work of Avery's group in the Hospital of the Institute, making this crowning achievement of experimental medicine a spectacular testimony to the explanatory power of the chemical view of life.

CHAPTER FOUR

AVERY'S PERSONAL LIFE

Private Life and Professional Life

The necrologies of its deceased members published by the National Academy of Sciences show that a large percentage of them came from rather humble homes, and that many were the sons of Protestant clergymen. In nineteenth-century America, a clergyman's way of life often seemed to provide his children with an ethical and cultural environment favorable for intellectual growth, leading eventually to membership in professional societies, including the Academy. This appears to have been true for Oswald Theodore Avery, who was born on October 21, 1877, in Halifax, Nova Scotia, four years after his parents had emigrated from England; his father was pastor of a Canadian Baptist church.

Although Avery loved to tell stories about himself, he avoided conversations of a purely personal nature, in particular those involving his family or the very early years of his life. He probably would have regarded any search into his familial background as an unjustified intrusion into his personal affairs; moreover, he would have felt that the information thus obtained could not possibly throw useful light on his scientific achievements. His attitude in this regard is apparent in the obituary he prepared for Karl Landsteiner, the somber genius who had been his colleague at The Rockefeller Institute for Medical Research.[1] The original text of the obituary that Avery submitted dealt exclusively with Landsteiner's scientific life, but the editor of the journal in which it was to be published requested that it be supplemented with details of the scientist's family life and behavioral peculiarities. Avery refused, with the statement that such personal details would not contribute to the understanding either of Landsteiner's scientific achievements or of his intellectual processes.

Claude Bernard had expressed similar views in *Le Cahier Rouge*, the notebook to which he confided his casual thoughts. "A great man is not great when he goes to bed, gets up, sneezes, etc., but only when he writes, thinks, and even then it is only on special occasions, as is the case for an actor. It is in these moments that man is truly great, and that we can reach him through his works. We had better ignore the rest; it does not add anything to the man."[2]

Claude Bernard and Avery may have been correct in believing that familial background and behavioral characteristics have little bearing on creativeness in science, the arts, or other intellectual pursuits. However, familial and behavioral factors inevitably influence a person's way of life, and thereby condition the manner in which creativeness is expressed, with regard to both form and content. Such conditioning can be illustrated by comparing the scientific careers of Avery and William Henry Welch. The qualitative difference in the contributions these two physicians made to biomedical sciences was not due to differences in their intellectual endowment, but to choices they made with regard to their ways of life — choices which probably originated from early familial influences and from temperamental peculiarities.

Both Welch and Avery studied medicine at the College of Physicians and Surgeons in New York City, where they received the best clinical training available at that time in the United States. Both did well in their academic studies, had great charm and skill in human relationships, and were judged by their teachers to have the attributes required for successful careers in the practice of medicine. Both, however, were dissatisfied with medical knowledge as it existed in their time, and abandoned clinical medicine as soon as they had the chance to devote themselves to laboratory research.

Although Welch and Avery took similar initial steps when they shifted their interest from the bedside to the laboratory, their subsequent courses were very different. Welch became more and more involved in medical education and statesmanship; Avery moved increasingly toward theoretical scientific work. Their temperamental characteristics certainly accounted in large part for this fundamental difference in their scientific evolutions.

Welch had a Gargantuan appetite, and was fond of ice cream and other sweets; he soon became obese. He loved the carnival aspects of life and to mix with crowds in Atlantic City and on the New York beaches. In a letter to his sister, written when he was over 50, he describes with gusto the excitement he experienced riding the roller coaster. With these popular tastes, it is not surprising that he found it easy to move into public life and to spend an enormous amount of energy lecturing, organizing medical groups, and engaging in medical or public health politics.[3]

In contrast, Avery ate very little, was extremely fastidious about the nature of his food, shunned public gatherings, and resented being entertained. Although he was a very effective lecturer, and loved to advise those who came to him, he virtually gave up public speaking after joining the research staff of the Institute. He kept shy of social responsibilities, and instead devoted all his energy and talent to laboratory work in collaboration with a small number of colleagues. Thus, Welch's extroverted personality led him to the creation of a social environment in which medical research became respectable and, indeed, fashionable, whereas Avery's introverted attitude enabled him to take full advantage of this environment to create new scientific knowledge.

Avery's natural endowments could certainly have enabled him to achieve great worldly success in any of several different fields, but he elected to withdraw almost completely from public life. He has left no written document to account for this choice, nor does he seem explicitly to have stated his reasons for it to either family or friend. His conversation was always sparkling and often penetrating, but he was very selective in what he revealed of his complex personality. To the end, he kept his own counsel. I apologize to his memory for trying to uncover in the following pages certain aspects of his personal life that he had chosen not to make public.

Familial Background

Avery's paternal grandfather, Joseph Henry Avery, was born and lived in England, where he was a papermaker in charge of paper manufacture for Oxford University. He must have had some inventive talent, as he was the first to make the thin paper that could be printed on both sides and used for the Oxford Bibles. On May 17, 1881, the delegates of the Clarendon Press at Oxford presented Joseph with a Bible "in acknowledgement of great services rendered by him to the Press during the publication of the Revised Version of the New Testament."[4]

Avery's father, Joseph Francis Avery (1846–1892), was born at Norwich, Norfolk. He had a mystical nature, and was not satisfied with the profession of papermaker. Early in his life, he came under the influence of a Baptist evangelist, C. H. Spurgeon, who was conducting a series of religious meetings in England. Although Joseph had been raised in the Church of England, he decided, on the basis of this experience, to prepare himself for the Baptist ministry. In 1870, he married Elizabeth Crowdy (1843–1910), who was three years his senior, and spent the first three years of his married life in pastoral service in England. Then he and his wife migrated to Canada, for somewhat obscure reasons. In his own words, "a strange impression took possession of the writer, – 'You are wanted and must go to Nova Scotia.' Against the advice of friends, including Rev. C. H. Spurgeon the desire and impulse grew; till in faith and not by sight, in

May, 1873, it was determined to break up the home and if needs be, risk and sacrifice everything and go not knowing whither, trusting in God's leading. Confident a church and work awaited on the other side."[5]

As the steamer landed, a welcoming committee was at the pier, and asked the Reverend Avery to preach on the following Sunday at the North Baptist Church in Halifax. He accepted, and remained as pastor for a year and a half, when "providentially the way opened to organize a new cause and church." The new Baptist church created by Joseph Francis Avery was called The Tabernacle, and he remained its pastor until 1887.

All indications are that both he and his wife were popular and successful in Halifax, yet they pulled up stakes once more when he received an invitation to be pastor of a Baptist mission church in New York City. Again, the reasons for the move are far from clear; a spirit of restlessness probably played some role, along with the divine call to duty:

> It did at first appear, and even now does sometimes seem painfully strange, that our home and church life should again be disturbed, just as the homestead began to yield its fruits, and the church by the establishment and growth of time offered prospective easement from the necessary toil which comes to the pioneer worker. But knowing it is always safe to give heed to the voice of God, we have listened, watched, and prayed, and now, fully persuaded the Master in His providence has called us to the greater city of New York, we are resolved to go forward . . . ; the thought and expectation was to spend and be spent in building the upper structure of the Tabernacle, but the builder made a delay of several weeks in getting out his estimates. Meantime an increasing desire for more spiritual and direct evangelistic effort grew, and by reading an article in the *Christian at Work* a strange agitation of soul was created. The facts and figures given showed how vast the field, how great the need of direct, patient, continual pastoral effort in pastoral work amongst the multitudes of New York.[6]

In 1887, J. F. Avery became pastor of the Mariners' Temple, situated on the lower East Side of New York City at 1 Henry Street, a section of the city that was notorious for its poverty and rowdyism. In the words of the Reverend Avery's wife:

> People, people everywhere. Crowded into the lofty tenement houses, burrowing in basements, packed in cheap lodging-houses, and swarming on the streets. To the casual observer the picture is bewildering. Even to the ordinary Christian worker the situation is one that would seem to defy all effort to improve it. Vice in a hundred repulsive forms holds many in its iron grasp. Relentless lust and passion hold captive many who long ago have lost the power to resist. Others are held in bondage

which if not so repulsive in its outward manifestation, is no less fatal in its final influence on human destiny — that of religious superstition.[7]

The squalor of the Bowery did not dismay the Baptist Averys, who accepted the challenge of having to deal with both Jews and Catholics, and with a neighborhood which was a melting pot of sin. They made the Mariners' Temple a lively center of religious and social activities until the Reverend Avery's death from Bright's disease in 1892.

The peregrinations of the Reverend Avery and his wife strongly suggest that they were enterprising persons, and this is confirmed by the wide range of their activities in both Halifax and New York. J. F. Avery wrote in the local newspapers about social problems; in 1876 he published an edifying pamphlet entitled "The Voyage of Life"; until his death, he edited a church paper, *Buds and Blossoms,* in which calls to worship and hymns provided the framework for discussions of community and family affairs. He must also have dabbled in medicine, or at least in pharmacy, as judged from the fact that he patented a preparation called "Avery's Auraline," which he claimed was useful for the "relief and cure of deafness, earaches and noises in the head." His wife entered into partnership with a certain Jane Caroline Irish to promote "Avery's Auraline" commercially, but the project failed.

The problems of daily life in New York were often difficult for the Averys, living as they did on a small pastoral salary and using some of it for the publication of *Buds and Blossoms.* Fortunately, the Baptist community of the greater New York area was tightly woven, and was always available for help and encouragement in times of trouble. This spirit of brotherhood is illustrated by the account published in *Buds and Blossoms* of the fire that destroyed the Avery home in December, 1890. All the community pitched in; expressions of sympathy and financial assistance came from as far away as North Tarrytown. In particular, Mr. John D. Rockefeller sent a friendly letter and enclosed a check for $100.

Mr. Rockefeller was deeply involved in all activities of the Baptist Church, and for this reason contributed now and then to the missionary program of the Mariners' Temple. In letters to him that are as flamboyant in style as in handwriting, the Reverend Avery suggested that he was in need of some financial help to convert the Jews and Catholics of the neighborhood. He also wished that the Mariners' Temple be made as appealing to the Bowery derelicts as were the dens of sin among which it was located. "The saloons are so brightening up around Chatham Square, I am jealous for the Old Temple; it begins to look weather beaten."[8] Or again, "I wish we could out vie with attractiveness the brilliant but soul cursing saloon."[9] While Mr. Rockefeller was much in favor of the Reverend Avery's efforts, he felt more at ease doing God's work through the Baptist Mission. Nevertheless, he continued to help the Mariners' Temple directly, as seen, for example, in a letter written to him by Mrs. Avery in 1893, one year after her husband's death.[10]

A letter from Mr. Rockefeller to the Reverend Avery, dated December 30, 1890, reveals the closeness of the New York Baptist community:

Rev. J. F. Avery,
#1 Henry St.
New York, NY
My dear Sir:
    I inclose herein a Christmas check for $50, for yourself and your dear family and wish you all a happy New Year.
    We have skating at my house, and it occurred to me, that as you moved down from the North, you might be skaters. Can you not all come around and join us tomorrow afternoon between four and six? You will find an entrance on either side of the house. Put your hand through the gate, and pull the bolt.
Yours very truly,
(signed)   John D. Rockefeller[11]

According to John D. Rockefeller, Jr., "Father was always an enthusiastic skater." He arranged that a yard in back of his house at 4 West 54th Street be converted into a basin that was flooded and used as a skating rink when the weather was cold enough.[12] There he invited his Baptist acquaintances, as well as churchmen and educators whose activities he valued. Those were the happy days when the richest man in the world could simply tell people to "put your hand through the gate, and pull the bolt" when he invited them to his home!

Photographs of Mrs. Avery show her to be a small, strong-willed person. As her son Oswald clearly resembled her physically, and perhaps to some extent temperamentally, it seems worth mentioning a few facts that suggest how she managed her life.

In New York, as well as in Halifax, she seems to have been the moving spirit in making her husband's church a social center for the Baptist community. She continued the mission work and the publication of *Buds and Blossoms* after her husband's death. Her religious beliefs were complemented by an earthy practical sense that led her to put pressure on the readers of *Buds and Blossoms* for payment of their dues: "Please send the amount due for your subscription at once; by so doing, I shall be relieved of much care and anxiety."[13]

Among the many ordeals that she overcame, she once had the odd experience of being considered dead for several hours. This happened in Halifax in 1882, when she was 39 years old. After a few days of fever, chills, and intense perspiration, she became so sick that, in the words of her husband, "The death dew stood upon the face." Her two boys, Ernest and Oswald, were called to her side as her end appeared near, and "she charged [them] to do good and be good before they retired to the parlor below and fell on their knees." She became completely unconscious. Eventually "at 2 a.m. the form was stiffened and chilled, the jaw had fallen. . . . 'It is all over; she is gone,' said the doctor, 'I may as well go home'." She was then prepared for the death linen she had carefully set aside for such an eventuality, but two hours later she called for help and said, "I have been dead, have I not? Yes, I remember, Jesus waved me back and said 'Not yet my child.' Oh! how disappointed I was." Soon she took a cup of tea, "asked for a biscuit, and heartily enjoyed the same." She lived for almost 30 years after this dramatic experience, but the story of her "death" remained in the family, and it may have so affected the young Oswald, who was five when he witnessed the event, that it played some role later when he selected medicine as a profession.

After her husband's death in 1892, Mrs. Avery worked with the Baptist City Mission Society, then located near the Manhattan Bridge. In her work, she was associated with a great variety of wealthy people, among whom were the Rockefellers, the Vanderbilts, and the Sloans. In particular, she was close to Emily Vanderbilt Sloan, who took an interest in the two surviving sons. These social contacts made it possible for the boys to spend some time on great estates (unidentified) in New York State. She eventually moved to 1202 Lexington Avenue, where Oswald lived while going to medical school; his Colgate roommate, William Parke, also lived there as a boarder during his law-school years.

While in Halifax, the Averys had three sons. The oldest, Ernest, seems to have had unusual intellectual gifts. "When but a toddler, he was fond of getting on a stool, arrayed in his father's white collar and tie, and from an open book preach to surroundings rather than to an audience. From the start in life he had a stronger brain than physique."[14] He died in 1892 of an undefined illness, perhaps tuberculosis; the account of his death in *Buds and Blossoms* reads like the hagiography of a medieval saint.

The youngest son, Roy, who was born in 1885, was also sickly during his early years. Much can be learned in *Buds and Blossoms* of his mother's struggle to protect his health, and he survived. As he was only six years old when the Reverend Avery died in 1892, "he did not remember him so that his brother [Oswald] eight years his senior was more of a father than a brother; he looked up to him and admired him greatly."[15] Roy followed his brother in the field of bacteriology, and eventually taught at Vanderbilt University Medical School in Nashville, Tennessee. Much of the information about the Avery family used in the present account was acquired by him and transmitted to me by his widow, Mrs. Catherine Avery.

Oswald, the second boy, was born in 1877. He was then referred to as "Ossie," but the nickname does not seem to have stuck with him long. Perhaps because his health did not generate as much concern as did that of his brothers, and more likely because he was, from the beginning, a very independent child, mentions of him in *Buds and Blossoms* are rather casual. Small in stature, he had a strong and extremely intelligent face. He began taking part in the activities of the church at a very early age, with the same kind of determination that was to serve him well later in his scientific work.[16]

When the Averys arrived in New York, the organ of the Mariners' Temple was in such poor condition that it could not be used, and there were no funds to replace or repair it. Enterprising as usual, Mrs. Avery managed to induce a young German musician to play his cornet in the church. Soon, her two oldest boys, Ernest and Oswald, took advantage of this new acquaintance to familiarize themselves with the cornet. Without help or prompting from anyone, they "got hold of an old and inferior instrument, and before we could believe they had, on the housetop, without raising any protest, both learned to play this somewhat difficult instrument."[17] The German cornetist was so impressed by the efforts of the two boys that he volunteered to give them free lessons, and within three months they were capable of playing with him in the church. When the young German returned to his homeland, he arranged for another cornetist to continue the musical education of the Avery boys. Eventually, they both became so proficient with the cornet that they obtained scholarships at the National Conservatory of Music.

Obtaining a good cornet was quite a problem for the Avery family. The first-class instrument recommended by a famous Boston cornetist cost more than $60, a large sum for a pastoral budget; but Ernest and Ossie would have none other because "it did not take nearly so much effort to blow, and produced a fuller, grander note."[18] Fortunately, friends of the Mariners' Temple became interested and contributed the necessary funds. As early as July, 1889, Ossie had used the new instrument in the temperance Sunday School and at the "regular meetings" of the church.

For a while, Ernest and Oswald made it a practice to stand on the steps of the Mariners' Temple on Sunday afternoon, playing their cornets to attract worshippers. The Reverend Avery mentioned this fact with pride in his letters to Mr. John D. Rockefeller, asking for financial help. In 1891, however, Ernest was so sick that he no longer had "the lung power for successful and continued playing."[19] Another cornetist had to be found to take his place, but Oswald continued until he left for Colgate Academy in 1893. Eventually, he became such an accomplished musician that he once played in Antonin Dvořák's Symphony No. 5, *From the New World,* with the National Academy of Music, under the direction of Walter Damrosch.

The Colgate Years

Oswald Avery was ten years old when his family moved from Halifax to New York City. Although he could not avoid coming into contact with the riff-raff of the city all around the Mariners' Temple, there is no indication that he was influenced in any way by this experience. He survived any vicissitudes which the neighborhood might have presented, and attended with success the New York Male Grammar School, from which he received a diploma in 1893.

He then moved to the Colgate Academy, and in 1896 entered Colgate University, from which he received the B.A. degree in 1900. He never referred to his early childhood, but he frequently spoke of his Colgate experiences, probably because the college years represented the beginning of a new phase of his life, during which he achieved intellectual independence from his familial background.

Colgate Academy and Colgate University, both in Hamilton, New York, had been founded in 1819 by the Baptist Education Society of the State of New York; a theological seminary was attached to the University.[20] The intellectual atmosphere of the school seems to have been extremely liberal at the end of the nineteenth century. Harry Emerson Fosdick, who was to become one of the most celebrated churchmen and preachers of America and who was Avery's classmate, has written entertainingly of the fact that his education at Colgate almost made him an agnostic by the end of his sophomore year. In his words, "wild horses could not have dragged me into church. . . . The old class prayer meetings saw me no more."[21] Although there is no information concerning Avery's religious attitude, it is probable that he, too, came to question some of his family's fundamentalist convictions.

The revolt against orthodoxy was very much in the air at Colgate at that time. A group of six students, among them Avery and Harry Emerson Fosdick, asked a young professor of philosophy to organize for them, during the senior year, a special course of metaphysics in order that they might examine the credibility of the Christian faith. One day, as the small group stood on the steps of Alumni Hall after class, Avery concluded their discussion with the startling pronouncement, "Fellows, you know there really *is* a God."[22] This from a boy who a very few years before had played the cornet to lure the unbelievers to conversion and who, in his adult life, made it a policy never to utter a statement that he could not document with overwhelming laboratory evidence!

Life was rugged at Colgate at the turn of the century. The students were expected to attend to their own housekeeping and to supply themselves with cloths, mop, broom, bucket, and dust pan. From the top of the hill where the college was situated, they had to walk down to the Hamilton village store for their supplies, especially to secure kerosene for their lamps; often the trip was made through unplowed snow.

Also during the winter, when the temperature hovered around zero outside, the fire in the little iron stove in the dormitory often went out during the night, and water froze in the pitchers. The students had to hustle downstairs with the scuttle of ashes and get coal and kindling from the pile outside the building. When the outdoor hand pump was frozen tight, the ice had to be melted with a twist of blazing newspaper, and water had to be found somewhere to prime the valve. Haste was essential, as breakfast and morning classes started at an early hour.

Small and elfish as he was, Oswald seems to have fared remarkably well under these rugged conditions. His schoolmates called him "Babe," probably because of his small size, but they also referred to his "aristocratic daintiness," which they traced to what they mysteriously termed "his residence among the dignitaries of the pie belt."[23] He continued to play the cornet (solo B flat) through his Colgate years and, in fact, became the leader of the college band. The photographs of him among the other members of the band show him to be small and slender of body, but with a face giving an impression of alertness, intensity, forcefulness, and a touch of youthful arrogance (Figure 4). During his junior year, it was said of him in the yearbook that only the accident of having been born in Halifax and therefore a foreigner prevented him from pursuing "his aspirations for the Presidency."[24] Parts of the yearbook characterization of Avery as a college student are noteworthy because they present such a sharp contrast with the adult man who became legendary in later years as a gentle, retiring, and seemingly shy scholar.

"Being a minister's son, he is blessed with a faith in Providence, *second only to his faith in himself.* . . . He lives in New York City, except in the summer which he spends with the scions of America's saponaceous aristocracy. He believes in the Baptist doctrine, Free Trade, American expansion and domestic finish"[25] (italics mine). The mention of his "residence among the dignitaries of the pie belt" and of the summers he spent "with the scions of America's saponaceous aristocracy" probably refer to the fact that, through his mother's social contacts, he had mixed with the well-bred, prosperous classes and had acquired some of their behavioral patterns. The "saponaceous aristocracy" included, in particular, Mr. J. P. Pyle, who marketed the Pearline and other kinds of soap, as well as the Pearline washtubs. Both Mr. and Mrs. Pyle were very active in the affairs of the Mariners' Temple. Their names appear repeatedly in *Buds and Blossoms,* both in church matters and because of their contributions to the familial life of the Averys. Judging from the statement in the Colgate yearbook, Oswald must have bragged about his social connection with such well-known and wealthy people.

From the beginning of his studies at Colgate Academy and University, Avery made excellent grades in all his courses. At the University, his average was 8.5 (out of 10) for the freshman year, and above 9 for the other three years; he majored in the humanities, and took only the few elementary courses in science that were compulsory.[26]

Paradoxical as it may seem for a person who later made it a point to avoid public appearances, his best grades at Colgate were in public speaking. Each of the four years he achieved grades of 9.5 in the subjects listed as Public Speaking, Oration, or Debate. At Colgate in those days, oratorical contests caused as much excitement as football games do today. On a famous occasion, the judges announced that there would be no second prize award; a tie for the first prize was to be shared by Harry Emerson Fosdick and none other than Oswald T. Avery.[27] Decades later, Avery was still prone to declaim in the laboratory, with obvious pleasure, the sonorous phrases of a speech on Chinese civilization that had been one of his college oratorical triumphs.

During the fourth year, which was then entirely elective, Avery took the following courses, in all of which he made excellent grades: Philosophy, Modern Philosophy, Ethics, History, English Literature, History of Art, Economics, Political Economy, and of course Public Speaking, with Debate for good measure. Many elective courses were offered in scientific subjects, but he did not choose to take any of them. This was the academic preparation with which he graduated from Colgate University on June 21, 1900. He entered the College of Physicians and Surgeons of Columbia University in New York City the same year.

Medical Education

It can be surmised that when Avery first went to Colgate, his intention, or at least that of his mother, had been for him to enter the ministry, as did many of his classmates; this might explain his interest in public speaking while at college. As mentioned earlier, his attitude in religious matters seems to have changed profoundly in the course of his college years. Like Harry Emerson Fosdick, he probably rebelled against "the kind of bibliolatry and theology" he had been taught. However, this alone does not explain why he chose to enter medical school after having emphasized philosophy, literature, and public speaking throughout his studies and, in particular, during his elective senior year.

A possible explanation might be the contacts he had had with problems of disease during his youth—his father's patent for ear ailments, his brother's death, his mother's near-death. However, personal tragedies were common at that time, and could hardly have been sufficient for Avery's decision. A more likely reason may be that medicine provided him with an outlet compatible with his familial missionary background and with the rational philosophy he had developed at Colgate (see Chapter Twelve).

Furthermore, medicine enjoyed great prestige around 1900, because recent spectacular discoveries in the field of infectious diseases opened possibilities for effective action in the future. In his autobiography, Harry Emerson Fosdick mentions that, after losing some of his original religious faith, he himself had considered going into medicine upon graduation from Colgate. According to the Colgate yearbook, four other students out of 30 in the class of 1900 expressed the intention to go to medical school. Three of them, including Avery, did, and they all went to the College of Physicians and Surgeons. At approximately the same time, the Reverend Gates read Osler's textbook of medicine and concluded from it that the furtherance of medical research would provide a worthwhile cause for Mr. Rockefeller's philanthropic interests (Chapter Two). Medicine apparently fitted well into the mood of the Baptist community at that time.

Whatever the reasons that led Avery to choose medicine as a profession, his deficiency in scientific training was not a handicap, as the scientific entrance requirements of the College of Physicians and Surgeons were virtually nonexistent. Courses in physics and chemistry were then part of the first-year curriculum. The only records of his medical education that have survived are course grades. These were good, except in bacteriology and pathology, the sciences to which he was later to make such monumental contributions! The nickname "Babe" had followed him from Colgate, and he was known by it during his four years of medical school. Dr.

Edwards A. Park (who became one of the foremost American pediatricians) had been his schoolmate and has stated to at least two persons that "Babe" was quite suitable to Avery as a medical student and a young doctor because he appeared so immature and the most unlikely to succeed.[28]

The College of Physicians and Surgeons had long been one of the leading medical schools in the United States but, at the turn of the century, it had not yet been much influenced by the scientific spirit. According to Alfred E. Cohn, who was a student there at the same time as Avery, the school was concerned almost exclusively "with the care of the sick." Since Avery never spoke of his medical school years, it is probable that they did not provide him with much intellectual satisfaction.

Immediately after receiving his medical degree in 1904, he joined a group engaged in the practice of "general surgery" in New York City. Around the turn of the century, this expression meant the general practice of medicine. One of the few existing testimonies of that period, if not the only one, is a thermometer in a silver case with the following inscription:

PRESENTED TO DR. O. T. AVERY BY
NEW YORK CITY TRAINING SCHOOL
JUNE 1, 1906

He remained in practice until approximately 1907, but found it upsetting to deal with patients suffering from chronic pulmonary disease and intractable asthma for whom he could do nothing really useful. From his own accounts, he was quite successful in his personal relations with patients, but clinical practice did not satisfy him intellectually and emotionally, probably for the reasons mentioned earlier. In the words of his close friend, Dr. A. R. Dochez, the experience "supplied him with some amusing stories but did not attract him sufficiently to make a career in that field."[29] Fortunately for him, medical New York was then becoming research-conscious, and he soon found an opportunity to shift from clinical to laboratory work. As this part of his life is the only one for which he himself has provided some documentation, it seems best to let him tell the story in his own words:

Sir Almroth Wright came to New York from England and gave a lecture at the Academy of Medicine on his newly invented opsonic technic. The New York City Health Department was interested in this and arranged to have a colleague of Sir Almroth give a short course of instruction to a small group. [Dr. Wright was the prototype of the physician in G.B. Shaw's play, *The Doctor's Dilemma*.]

I was one of those to take this course. At its completion, Dr. William

Park gave me a job doing opsonic indices for the Board of Health at a stipend of $50 per month for part-time work. I also found part-time employment doing milk bacteriology for the Sheffield Company. Pasteurization of milk was just coming in; I made bacterial counts of milk before and after pasteurization at a stipend that was also $50 per month.[30]

The next important move in his professional life was his appointment to the Hoagland Laboratory in Brooklyn, an institution which, as mentioned earlier, has historical interest because it was the first privately endowed laboratory for bacteriological research in the United States. The director, Benjamin White, was not a physician. In 1903, he had earned a Ph.D. in physiological chemistry at the Sheffield Scientific School of Yale University and progressively acquired practical knowledge of medical microbiology, first in the United States, then in Germany, Austria, and London. When he took over at Hoagland in 1907, his first administrative act was to appoint Avery to the position of associate director at a salary of $1,200, which was increased to $1,500 in 1909. This is how it happened, again in Avery's words:

Benjamin White and I met in this way. While I was a student at the College of Physicians and Surgeons, I roomed with a young law student, William M. Parke, who after admission to the bar practiced for a time on Remsen Street. It so happened that Benjamin White lived in the same house, and thus we became acquainted. White mentioned to me that he needed a young doctor to be his assistant director. I responded enthusiastically and so I was invited over to the Hoagland Laboratory.[31]

It will be remembered that Parke had been Avery's classmate at Colgate Academy and his roommate at Colgate University. While studying at the New York Law School, he roomed in the apartment occupied on 1202 Lexington Avenue by Avery and his mother.

The six years Avery spent at the Hoagland Laboratory were of crucial importance for his scientific development. Benjamin White, having been trained as a chemist, could indoctrinate him in laboratory techniques and in the chemical mode of thinking. Moreover, the responsibilities of the department were wide, with regard to both research and teaching, thus providing him with a highly diversified experience. He and White decided at the outset that they would treat all bacterial cultures as if they were plague bacilli. This set the stage for the exceedingly careful techniques that characterized Avery throughout his professional life.

The first problem Avery had to deal with was the bacteriology of yogurt and other fermented milks that were just becoming popular through the work of Elie Metchnikoff at the Pasteur Institute in Paris. Metchnikoff claimed that the consumption of these fermented milk products accounted for the great longevity of populations in Eastern Europe, because they prevented intestinal intoxication by controlling the putrefying bacteria of the gut. As the Hoagland Laboratory was in a Syrian neighborhood where the grocers prepared their own fermented milk, Avery developed a taste for this product while studying it. He and White eventually recorded their findings in a paper entitled "Observations on Certain Lactic Acid Bacteria of the Bulgaricus Type." From 1909 to 1913, they carried out studies on a wide variety of medical problems, which they approached by optical, bacteriological, immunological, and chemical techniques. The subjects ranged from the demonstration of *Treponema pallidum* in syphilitic lesions to the analysis of the antigenic properties of certain plant proteins. Avery thus received, during these few years, a very broad practical training in various fields of bacteriology and immunology.

In 1910, White had a severe reactivation of tuberculosis, and went to the Trudeau Sanatorium in Saranac Lake in the Adirondacks to take the cure. Avery accompanied him on the initial trip, and later spent several periods of vacation at the sanatorium. This experience naturally stimulated in him an interest in tuberculosis, which he satisfied by working in the Trudeau laboratory and library. His notebooks of the time were full of extensive and carefully handwritten analyses of current publications on the clinical and experimental aspects of tuberculosis.

Avery's publications during the Hoagland Laboratory period were scholarly in approach and thorough in execution, but they exhibit little originality and can be regarded as the products of a self-training period. However, one of them deserves mention because it deals with a type of systematic clinical testing which was of lasting practical value, but was very different in research style from the more imaginative studies for which he was to become famous a few years later. While "vacationing" at the Trudeau sanatorium, he carried out 100 consecutive blood cultures of tuberculous patients in the active phase of their disease without ever recovering tubercle bacilli or observing evidence of secondary infection. These negative findings were important for the understanding of tuberculosis, and they demonstrate his ability to carry out a systematic clinical investigation. Although routine work of this kind was not his bent, it was fortunate that he undertook it, because it caught the attention of Dr. Rufus Cole, director of The Rockefeller Institute Hospital, and thus indirectly led Avery into the scientific environment best suited to the unfolding of his genius.

Some two years later, he carried out, in collaboration with Benjamin

White, a chemical and toxicological study of a product derived from tubercle bacilli by extraction with alkaline ethanol. This investigation, which was published in 1912, also was important in Avery's scientific development. It was the beginning of a pattern that can be recognized throughout his subsequent career at The Rockefeller Institute—the systematic effort to understand the biological activities of pathogenic bacteria through a knowledge of their chemical composition. Another phase of his training took place in 1911, when he spent his vacation at the biological laboratories of the H. K. Mulford Company, instructing its staff in bacteriological techniques and learning from them industrial methods for the production of antitoxins and vaccines. This practical experience served him well two years later, when he was made responsible for the production of antipneumococcus therapeutic sera at the Institute.

Avery published nine papers during his Hoagland Laboratory period, one of them a chapter on "Opsonins and Vaccine Therapy" that he prepared in collaboration with Dr. N. B. Potter for Hare's *Modern Treatment,* a text of clinical medicine that was then widely used. Here, again, this publication contributed to his scientific career, because it prepared him for the study of the role of phagocytosis in infectious processes.

The Hoagland Laboratory experience also provided Avery with the chance to use in medicine the expository gifts he had displayed at Colgate University. In collaboration with White, he worked on the bacteriology of postsurgical infections and even planned to write a monograph on the topic. To obtain material for this project, he encouraged clinicians of the Brooklyn area to bring their bacteriological problems to his laboratory and, when they did so, he gave them elaborate individual advice. Thus began his practice of teaching by conversation, which he employed later with great success at the Institute.

Another of his responsibilities at the Hoagland Laboratory was to run a course for the student nurses. To impress the students with the dangers of conveying respiratory germs by sneezing, he told them, "If your saliva were blue, you would have to look at your patients through a blue fog." From then on he was referred to as "The Professor" and later, more familiarly, as "Fess," not only because of his skill as an expositor of science, but also because of his wisdom in counsel.

The study on secondary infections in pulmonary tuberculosis, referred to earlier, which Avery published in 1913, had greatly impressed Dr. Rufus Cole. Late in the spring of that year, Cole paid a seemingly casual visit to the Hoagland Laboratory, where he found Avery working with cultures of pneumococcus and testing their solubility in bile. He engaged the conversation with him by pointing out that, at the Institute, they carried out the solubility test with buffered solutions of pure bile salts, rather than with crude bile. The discussion that followed convinced Cole that Avery had the proper scientific qualifications for the pneumonia research program at the Hospital. This program involved a comparative study of the different pneumococcal types, and could best be approached by a scientist with knowledge of bacteriology, immunology, and chemistry. This was precisely the unusual combination of skills that Avery had acquired at the Hoagland Laboratory, under the guidance of White.

A few days after this meeting, Avery was invited to visit The Rockefeller Institute, where he had lunch with Dr. Flexner, who also must have been impressed. Shortly after, he received a letter from Dr. Cole with the offer of a position as bacteriologist to the Hospital. In view of the prestige of The Rockefeller Institute, and of the invitation to participate in the program on lobar pneumonia, a disease of which his mother died in 1910, he was certainly interested in the offer. However, according to his own account, he did not reply, in part out of negligence and in part because he was not eager to change positions. He liked his colleagues at the Hoagland Laboratory and especially enjoyed the complete freedom to work at what he wanted without ever being put under pressure. Dr. Cole wrote a second letter and, receiving no reply, drove once more to Brooklyn, almost apologizing to Avery for having offered him a position with inadequate salary; the purpose of the second visit was to offer more attractive conditions. Avery accepted, and joined the Hospital of the Institute in September, 1913. Only later did Dr. Cole realize that Avery tended to neglect his correspondence, and that he had ignored the initial offer not because it was financially inadequate, but because he had more urgent and more interesting things to do than to acknowledge a business letter.

The Rockefeller Institute Years

Avery joined the Institute with the title of Assistant; he was promoted to Associate in 1915, to Associate Member in 1919, and to full Membership in 1923. He became Emeritus Member upon retirement in 1943 at age 65, but continued working in his laboratory until 1948. Since the scientific aspects of his New York phase will be presented at length in subsequent chapters, it will suffice to outline here some facts of his private life.

Shortly after his arrival in New York, Avery began to share an apartment with Dr. A. R. Dochez, who was then his colleague in the department of respiratory diseases at the Hospital and who was also a bachelor.

They continued this arrangement even after Dochez became Professor of Medicine at the College of Physicians and Surgeons. As their apartment was rather large, they took with them, for various periods of time, other young medical scientists who were not yet married. In these early days, their furniture consisted chiefly of odds and ends brought in and abandoned by each of the successive occupants.

In 1927, Avery and Dochez moved to 67th Street, between Lexington and Third avenues, directly across the street from the fire station and police precinct (the site is now occupied by a new building that houses the Russian Embassy). They were joined at that time by one of Dr. Dochez's brothers, who was a businessman and a widower and who brought with him a great deal of fine household furnishings. This is the way Avery reported the event in a letter to his sister-in-law, dated October 30, 1927:

> . . . we have now a really delightfully equipped apartment with some beautiful pieces — ranging from genuine Chippendale, original color prints, and oriental rugs to Worcester Royal China and massive silver service. It's really a great treat after years of association with an ill-assortment of golden oak furniture and non-descript iron bedsteads of the Early Wanamaker-Grand Rapids period.

When Dochez's brother remarried several years later, he took away all the valuable pieces he had contributed, compelling the two bachelors to fill the gaps with what they could find and afford at auction sales during the depression years.

The Avery-Dochez establishment was managed by Elsa, a Danish housekeeper whose jovial mood and wholesome food made the apartment a comfortable and carefree home. When the household was finally broken up in 1948, Dochez took only a few items to the single room in which he settled at the University Club in New York, and Avery moved the rest of the furniture to the house he rented in Nashville, Tennessee.

The outward manifestations of Avery's life in New York were extremely simple and frugal. Every day, just before 9 A.M., he walked the few blocks from his 67th Street residence to the Hospital at 66th Street and York Avenue. As soon as he reached his office on the sixth floor, he shed his subdued gray jacket for an equally subdued light-tan laboratory coat. He then took position at his desk, where one or several of us soon joined him to begin the day with conversations, the tone of which I shall describe in the next chapter. On special occasions, he put on a white laboratory coat instead of a tan one, for example, when he had to call on Dr. Cole or Dr. Flexner for some administrative problem, and every Wednesday morning for the so-called "ward rounds," which, in reality, were held in the Hospital solarium. Every day, he went down for lunch in the Institute dining room; on the first and third Monday of each month he faithfully attended the dinner of the Hospital journal club. While he watched carefully everything that was going on around him at these gatherings, he seldom volunteered to participate in the discussions that took place. At lunch, at ward rounds, or at the journal club, he talked only when asked for his opinion, and even then his answers were short and to the point. The great skill in public debate he had displayed at Colgate University never expressed itself on The Rockefeller Institute campus.

He was immensely popular among colleagues and outsiders, both men and women. Many were those who were eager to entertain him, and he could have spent every evening out if he had so desired; but, in fact, his social life was extremely restricted. Outside his laboratory, his most enjoyable moments appear to have been when Dochez returned home late at night from one of his countless social engagements. Then talk would begin on almost any topic, but preferably on one related to medical science and to the theoretical problems of infectious diseases. Not infrequently, when Dochez returned from the Metropolitan Opera, he found Avery reading quietly in bed. Then he "would sit down in full evening dress and with great animation describe to his old friend some of the illuminating thoughts on the subject of microbiology which had occurred to him during the second act of *La Traviata,* or whatever the evening's opera had been."[32]

Both Avery and Dochez claimed that they derived much knowledge from these midnight discussions. However, it is likely that the most profitable result of their interplay was not what they learned from each other, but that they used each other as perceptive sounding boards, better to define whatever question each had in mind. These midnight discussions sharpened their thoughts and gave them a form that could be successfully communicated to other listeners and converted subsequently into laboratory tests.

The picture I have just drawn may give the impression that Avery's life was rather drab, all work and no play. In reality, he managed to enliven every moment of it with subtle attitudes and remarks that made his company a pure delight. Dr. Colin MacLeod, who was one of his closest associates around 1940, has given an amusing account of a typical afternoon in 1941, while Avery was preparing the speech he had to deliver in the month of May as president of the Society of American Bacteriologists:

We talked about whether he should say that bacteriology is the "Queen of the Biological Sciences," or, as I might suggest, the Crown Princess, because she hadn't arrived yet; and so we spent the last half hour of the late afternoon, until Fess would say, "Let's go and see Do" [Dr. Dochez]. And then a short three and a half block walk across town to

see Do, who would greet us, rubbing his hands and saying with enthusiasm, "Hey, you're late, Fess. I'll make a Martini," which he would do
forthwith, and when brought, would exclaim "Fess, drink it up before
the bloom goes off it!"

And so then an extraordinary hour out of many with these two
wonderful gentlemen — bachelors — who knew about the goodness of life
and of science and complemented each other in a way I have never seen
elsewhere.

We might end up on this occasion declaring that bacteriology was not
a Crown Princess or even a Cinderella, but more likely a pumpkin. But
you can be sure we had a stimulating time.[33]

The even tenure of Avery's life was disturbed during the early 1930s
when he suffered from Graves' disease. He then frequently experienced
moods of depression and of irritation that he did not always manage to
conceal, despite valiant efforts. Finally, he underwent a thyroidectomy at
the Presbyterian Hospital in New York (either in 1933 or 1934 — the
hospital records have been destroyed), and recuperated for several months
at the residence of his friend Dr. Harry Bray, who was superintendent of
the Raybrook Sanatorium in the Adirondacks. When he returned to the
Institute in the fall, he was once more his old self, but took advantage of his
medical condition to decrease still further his social commitments and
devote himself more completely to his work and to his departmental
associates.

Early during his New York years, he began to spend his vacation by the
seashore, and developed a great love of sailing. At first he went to
Gloucester, Massachusetts, where he was taken by his friend and colleague
Dr. Homer Swift, who owned property there. He rented a house on the
edge of the city at Stage Fort Park, which he shared with his brother's
family during the summer. While in Gloucester in 1929, he was invited by
Dr. Alan Chesney, who also had been his colleague at the Institute and was
then at The Johns Hopkins Medical School, to visit him on Deer Isle in
Maine. He immediately fell in love with the place, and from then on spent
every summer on Deer Isle, where several of his friends in scientific
medicine also spent their vacations.

One of his great pleasures was to go sailing on Penobscot Bay on Dr.
Chesney's sloop. According to Dr. Chesney, "He never really tried to
master the art, but . . . rarely missed an opportunity to go for an afternoon
sail when the occasion offered. Short in stature and small in body as he
was — he could scarcely have weighed much over a hundred pounds — one
could not imagine him ever participating in any competitive sport."[34] In
addition to sailing, he walked through the woods collecting ferns and

wildflowers; he observed the rapid growth and decay of toadstools, and wondered at the enzymatic mechanisms involved in these biological processes; he painted landscapes and seascapes in watercolors of a subtle and rather individual style.

Except for his annual summer trips to Gloucester and, later, to Deer Isle, Avery did very little traveling. In 1932, he went to San Francisco to deliver an important lecture, and came back via the Yosemite Valley, Los Angeles, and the Grand Canyon. Upon his return, he wrote to his brother that the trip had been "an unforgettable experience" and that he marveled "at the gigantic sculpturing of Nature," but he never repeated the experience and never referred to it again. For a scientist of his fame, he attended surprisingly few scientific meetings, even within the United States. He did not go to Germany when he was awarded the Paul Ehrlich Gold Medal in 1933, or to England when he was proposed for a doctorate *honoris causa* by Cambridge University in 1944 and awarded the Copley Medal by the Royal Society in 1945, or to Sweden when he was awarded the Pasteur Gold Medal by the Swedish Medical Society in 1950. He gave several excuses for not traveling: lack of time, poor health, or financial cost, but the only valid reason was that he restricted more and more the range of his experiences to what he could find in his laboratory work, on Deer Isle, and among his brother's family. In 1948, he decided that he had shot his bolt; as he no longer felt able to function effectively in the scientific arena, he retired to Nashville, Tennessee.

The Nashville Years

Avery's reason for moving to Nashville was that he would find there his brother Roy, who taught bacteriology at Vanderbilt University School of Medicine, his sister-in-law Catherine, his niece Margaret, and his cousin Minnie Wandell. Furthermore, he had several friends at the medical school, in particular Dr. Ernest Goodpasture, chairman of the department of pathology, and Dr. Hugh Morgan, chairman of the department of medicine, who had once worked in Avery's laboratory at The Rockefeller Institute.

Dr. Morgan persuaded Avery to continue with some laboratory work in Nashville, and arranged that he be given a research grant by the Department of Defense for the study of immunity to streptococcal infection. He also arranged that Dr. Bertram E. Sprofkin, who had just completed his medical residency at The Johns Hopkins, join Avery for two years as a co-investigator. This program resulted in a joint report entitled "Studies on the bacteriolytic properties of *Streptomyces albus* and its action on hemolytic streptococci." According to Dr. Sprofkin, Avery often referred to the work with which he had been associated at Rockefeller, and "his enthusiasm for any information concerning the nucleic acids remained at a high level until his final illness."[35] On the whole, however, Avery made little effort to take advantage of the scientific facilities made available to him.

Throughout his years in New York, Avery always maintained contact with his family, but almost exclusively by correspondence, except during the summer vacations in Gloucester and on Deer Isle. In Nashville, however, the familial atmosphere became much more intimate. He was able to rent a fine stone house belonging to one of Roy's friends who had moved to a farm in the country. The house was set on more than one acre of land with beautiful trees, and had the additional advantage of being situated but a few doors away from Roy's own home. His cousin Minnie Wandell, "who adored him," acted as his housekeeper.

Nearly every night Roy would walk up the street to join him for a game of backgammon. It was a common sight to "see the two walking together from one house to the other, obviously enjoying to be together."[36] There is no doubt that the possibility of close associations with his sister-in-law and his niece provided him with the kind of emotional satisfaction from which he had been deprived by his way of life in New York.

He became very much a part of his neighborhood, where he was known not as a scientist but "as a very pleasant person to have around."[37] Since he had not previously lived in the South, he took great interest in the local flowers and trees, learning their names and peculiarities. "His appreciation of the flowers I shared with him from time to time would have warmed the heart of any gardener."[38] All accounts agree in giving the impression that, during the last years of his life, Avery managed to create around himself the atmosphere of the country gentleman. In fact, Dr. Sprofkin, who had not known him before, felt that Avery spoke in Nashville with a slight British accent, even though he had left Nova Scotia when he was ten years old and had never returned. At times, Avery expressed the opinion that "there was nothing so fine as a genuine British gentleman," and Dr. Sprofkin felt that one of the reasons he loved his Nashville stone house was that it embodied many of the most attractive features of an English cottage.

While on Deer Isle during the summer of 1954, he noticed discomfort in his right upper quadrant, and was examined first at The Rockefeller Hospital and then in Nashville. The initial tentative diagnosis was gallbladder disease, but surgery revealed extensive hepatoma (cancer of the liver). His terminal illness was very painful, but he bore it with his characteristic patience. He died at the age of 78 on February 20, 1955, and was buried in Mount Olivet cemetery in Nashville.

CHAPTER FIVE

AVERY'S LIFE
IN THE LABORATORY

The Inwardness of Research

Avery was a late starter in science. His research at the Hoagland Laboratory had been thorough and diversified, but neither path-breaking nor even intellectually adventurous. He was almost 36 years old when he was appointed to the staff of The Rockefeller Institute. The four papers he published in 1915 dealt with the application of conventional serological techniques to the biological classification of pneumococci isolated from patients, and were of a routine nature. He was probably then regarded as a competent medical bacteriologist, rather than as a creative scientist. In 1916, when he was 39 years old, there was nothing in his professional achievements to indicate that, from the age of 40 to the age of 65, he would continuously make major contributions to the biomedical sciences.

Graphologists, however, might have recognized in his handwriting unusual characteristics suggesting that he would go far if circumstances favored him. Figure 9 represents a handwritten bacteriological report that he prepared in 1916 on a pneumococcus culture isolated from a patient at The Rockefeller Institute Hospital. Figure 10 is a letter to Simon Flexner, written when Avery was ill with Graves' disease. In both cases, the script reveals aspects of his temperament that could hardly be guessed at from photographs of him taken during his adult life, or from descriptions of his usual social behavior. The flourish of the script suggests enthusiasm, versatility, and tenacity, a bold and imaginative mind, a love of form and fantasy, an affirmative and almost daring self-confidence. These attributes were concealed or muted in his public appearances, but they were frequently expressed in his laboratory life and became evident in his creative work after 1916. The bacteriological report is of historical interest for another reason. The pneumococcus culture which it describes, labeled D39, was widely used later by Avery and his collaborators in many phases of their research program; in particular, it gave rise to the strains used in the studies of transformation of pneumococcal types that led to the demonstration that deoxyribonucleic acid (DNA) is the carrier of genetic information (Chapter Eleven).

The affirmative and almost exuberant character of Avery's script symbolizes the boldness he began to display in his research style after joining the Institute. Whereas his earlier work had been conventional, the papers he published in collaboration with Dochez, first in 1916 on antiblastic immunity, then in 1917 on the soluble specific substances of pneumococci, describe approaches to immunological problems that were then entirely new — as original in execution as they were adventurous in interpretation.

One might assume that the profound change in Avery's research style that began in 1916 was simply the consequence of his being provided with generous budgets and elaborate resources for experimental work, but this is not the case. As we shall see later, Avery never had a large laboratory at the Institute, and he was always extremely frugal in the use of his research facilities. He rapidly developed into a creative scientist not because he was provided with funds and technical help, but because the Institute Hospital provided an intellectual and human atmosphere that suited his temperament.

The type of scientific environment Avery found in the Hospital, and his ideal of how biomedical research should be conducted, are lucidly expressed in the words he used in 1949 when he presented the Kober Medal of the Association of American Physicians to Dochez. Both Avery and Dochez were past retirement age at the time of the ceremony, but both had exemplified throughout their professional lives the thoughtful and parsimonious attitude that Avery ascribed to his friend in the following words:

> Throughout his [Dochez's] studies there is unique continuity of thought centering in the dominant problem of acute respiratory diseases. The results of his work are not random products of chance observation. They are the fruits of years of wise reflection, objective thinking and thoughtful experimentation. I have never seen his laboratory desk piled high with Petri dishes and bristling with test tubes like a forest wherein the trail ends and the searcher becomes lost in dense thickets of confused thought. I have never seen him so busy taking something out of one tube and putting it into another there was no time to think of why he was doing it or of what he was actually looking for. I have never known him to engage in purposeless rivalries or competitive research. But often I have seen him sit calmly by, lost in thought, while all around him others with great show of activity were flitting about like particles in Brownian motion; then, I have watched him rouse himself, smilingly saunter to his desk, assemble a few pipettes, borrow a few tubes of media, perhaps a jar of mice, and then do a simple experiment which answered the very question he had been thinking about when others thought he had been idling in aimless leisure.[1]

Avery himself possessed to an extreme degree the qualities he attributed to Dochez. He exemplified the attitude he liked to call "the inwardness of research" — a phrase he borrowed from Theobald Smith — to denote that scientific research implies both the establishment of facts through the trained senses and the processing of these facts through the inner cogitations of the intellect. He was scornful of ill-thought-out, busybody experimentation, the kind he was wont to describe with a gentle smile as taking something out of one test tube and putting it into another.

Whatever the importance or urgency of a problem, he never hurried, because he believed that worthwhile answers could come only from orderly thought based on careful observation and intellectual analysis. He was prone to convey the importance of observing small details by quoting the words of an old black patient who watched, with amused surprise, the young doctors rushing about the wards of The Johns Hopkins Hospital: "What's your hurry, Doc? By rushing that way, you passes by much more than you catches up with." One of the reasons Avery was so effective in his research is that he did not try to save time by being falsely efficient. He knew that mechanical efficiency is not the same as effectiveness.

He conducted his investigations with the least possible expenditure of physical effort, and with strict economy of materials, laboratory equipment, and experimental animals. For him, the ideal experiment was one that yielded a clear and inescapable conclusion from a limited number of facts observed in a few test tubes or a few animals.

Avery had read widely and deeply into the literature of experimental medicine before he joined the Institute. From the time he committed himself to the study of respiratory diseases, however, he made little effort to keep up with the details of other fields of science, let alone with other intellectual disciplines. I worked in his department during a stage in my life when I was under the illusion that one could assimilate the whole body of biological sciences. I was often surprised, and at times almost shocked, by the fact that his range of scientific information was not as broad as could have been assumed from his fame and from the variety and magnitude of his scientific achievements. Furthermore, his imagination did not seem to me of the kind that soars far above the concrete facts revealed by straightforward observation or by simple laboratory experiments. I now realize that these characteristics, which I regarded at the time as limitations in his scholarship and imaginative power, were in reality great assets from the point of view of his scientific creativity. Wordsworth's lines, "Wisdom is ofttimes nearer when we stoop than when we soar," fits well the manner in which Avery's imagination was intimately linked to the facts that he knew from *direct* experience. He did not indulge in vague, sweeping generalizations, but he had an uncanny gift for transmuting the details he had observed into an image of reality.

He had no taste for broad but shallow learning, and did not make pretense to knowledge unless he had made it a constituent part of his own intellectual fabric by using it in a creative way. For this reason, he tended to focus his reading on the publications that were related to the experimental work in which he was engaged at any given time; in fact, he almost limited his scientific contacts to those he could integrate into his own research program. But the extent and thoroughness of his search for information was truly phenomenal, once he had become committed to a particular project. He assimilated abstruse aspects of organic chemistry during the years he was involved in immunological studies with synthetic antigens, and he became familiar with theoretical genetics when, late in his professional life, he started the work that led to the identification of the substance responsible for hereditary transformations in pneumococci.

He studied carefully the books and articles that had a direct bearing on his problems, but he learned even more from personal contacts with anyone who could provide him with theoretical knowledge or practical information. I shall come back later to his skill in using conversation as an educational process, but I want to emphasize at this point that it was through his eagerness to learn from others that he developed a scientific staff characterized by great diversity of professional specialization and, at the same time, by a remarkable unity of purpose.

Picking other People's Brains

Avery never referred to his collaborators as assistants, or even as associates. When he wanted to mention the scientists, young or old, who had participated in his research programs, he used circumlocutions to indicate that they had been his friends, not subordinates — "the boys" or "the people who have been in this laboratory." In his acceptance speech for the Kober Medal in 1946, he gave credit for the success of his department to the inspiration and wisdom that had been provided by Rufus Cole, and he acknowledged the contributions of his associates with a characteristic understatement: "Cole picked these men, and all I had to do was pick their brains."[2] It is indeed true that he used our technical skills and derived many of his ideas from the theoretical knowledge and the practical know-how of our diversified scientific disciplines. The more interesting and important truth, however, is that it was he who formulated the objectives of our collective enterprise and also set our very research style.

The following two examples will illustrate how he "picked the brains" of his collaborators and how he managed to integrate their specialized contributions into his departmental program while helping each of us to discover and develop a personal scientific identity.

In their early days at the Hospital, Dochez and Avery had prepared, from broth cultures of the various pneumococcal types, crude fractions of soluble materials that were specific for the particular type of pneumococcus from which the material had been obtained. They referred to these crude fractions as specific soluble substances (SSS), each of which is specific for a particular pneumococcus type. They soon realized that it would be of great importance to establish the chemical nature of these substances, because this would explain the mechanisms of specificity and throw light on the pathological behavior of pneumococci. Avery's first great contribution to science was to devote himself to this problem, even though he did not have the knowledge of organic chemistry required for the isolation and identification of the specific soluble substances.

Using simple techniques that he liked to refer to as "kitchen chemistry," he managed to prepare and purify small amounts of the pneumococcal substances that were endowed with specific immunological activity. In the hope of finding a colleague who could help him to further the tasks of purification and chemical identification, he constantly carried in his pocket a small tube containing some of the mysterious powder. He had particularly in mind the organic chemist Michael Heidelberger, who was then working in the department of kidney disease on the seventh floor of the Hospital. The following account is one that I heard many a time from Avery, and that Dr. Heidelberger recently confirmed in all its details.

Every time he had a chance, Avery would agitate the tube of SSS in front of Heidelberger and say, "Michael, the whole secret of bacterial specificity is in this little tube. When can you work on it?" And Heidelberger would answer, "Fess, this is a very interesting problem, but I have to spend all my time making crystalline oxyhemoglobin. I shall look into your problem when I have succeeded in obtaining for Van Slyke's team good crystals of oxyhemoglobin." The scene repeated itself time and time again, but finally, out of interest and friendliness, Heidelberger was able to work on the specific soluble substance. Thus began a collaboration which eventually brought Heidelberger to join forces with Avery. He soon identified the active material of SSS as a polysaccharide, and became thereby one of the pioneers and great masters of immunochemistry.

The second example concerns the circumstances that resulted in my own association with Avery, beginning in 1927. I was then a graduate student in soil microbiology and soil chemistry at the New Jersey Agricultural Experiment Station. Through a series of accidents, I found myself, during the early spring of 1927, seated next to Avery in the old dining room of The Rockefeller Institute for Medical Research. I knew nothing of his work and, of course, he knew nothing of me. With his usual graciousness, however, he inquired about my scientific interests and about the topic of my Ph.D. thesis. I told him that I had been working on the microbial decomposition of cellulose in soil, and had isolated several species of bacteria and fungi that could destroy that substance. He immediately became intensely interested, and invited me to continue the conversation in his small office, where he asked for further details about my work. Then he began slowly to suggest that my bacteriological studies with cellulose were related to his own work with pneumococci. As I knew hardly anything about medical microbiology, he patiently explained that these microorganisms owe their virulence to the fact that they are protected against the defense mechanisms of the body by a mucilaginous envelope — the pneumococcal capsule. This capsule, he told me, is made up of a polysaccharide, a hemicelluloselike substance chemically related to the true cellulose that I was using in my own experiments. And then, as if by a casual gesture, but in fact deliberately, he took from the right-hand drawer of his desk a little tube containing a white powder, labeled in his neat handwriting SSS III [Specific Soluble Substance of type III pneumococcus] and shook it in front of me. Several years later he gave me this tube, still containing some of the SSS III, and I have kept it ever since as a talisman.

While shaking the tube, Avery said, "This is the polysaccharide of which the capsule is made. It is completely resistant to the body enzymes and to all the other enzymes we have used. It can be decomposed only by strong acid treatment. If only we knew of a way to decompose it with an agent mild enough to be used in the body — an enzyme, for example — much could be learned about pneumococcal infections." Even though I did not understand all the details of the problem, I was fascinated by it and, probably even more, by the scientific drama that emerged from Avery's words and from the quiet intensity of his gestures and facial expressions. Under the spell of his charm and contagious enthusiasm, I stated that, in my opinion, it was possible to discover such an enzyme. I outlined how this could be done, and even mentioned that I might find time to work on the problem at the end of the summer.

In the course of that very same afternoon, Avery introduced me to two persons whom I did not know; one was Rufus Cole, the other was Simon Flexner. Nothing was said by either of these gentlemen that I can remember, except for some general remarks about French bacteriologists and the Pasteur Institute. I went back to the New Jersey Agricultural Experiment Station and then traveled through the United States during the summer, as I had planned. While in Fargo, North Dakota, I received a telegram informing me that I had been granted a fellowship at The Rockefeller Institute for Medical Research, to work with Avery. I had not applied either for a job or a fellowship, but Avery had sensed that I could be of help to his work and, without even corresponding with me, he had taken the necessary steps for my appointment.

I joined his laboratory in September, 1927, and began to work on a topic of pneumococcal physiology that had caught my interest, but that had no relation to the search for an enzyme capable of decomposing the polysaccharide. Now and then, Avery gently reminded me of the original problem, and I finally began working on it during the summer of 1928. I obtained active preparations of a bacterial enzyme early in the summer of 1929, while Avery was vacationing in Maine, and immediately wrote him of my success. I mentioned especially that the enzyme had proved capable not only of decomposing the capsular polysaccharide *in vitro,* but also of destroying the capsules of pneumococci *in vivo*, thereby curing mice suffering from experimental pneumococcal infection. Avery immediately returned from Maine and together we repeated the experiments. The findings were published in 1929.

I have told this story in detail to define the parts played by Avery and by myself in this particular problem. There is no doubt that I contributed the idea of how to discover a soil bacterium capable of decomposing the capsular polysaccharide of type III pneumococcus; I also worked out the techniques for the isolation of the bacterium, for making it produce large amounts of the enzyme, for extracting and purifying the enzyme, and for testing its activity in both the test tube and animals. But there is another side of the story.

Now that I have read Avery's reports to the Board of Scientific Directors of the Institute, I know that, as early as 1923, he and Heidelberger had tried without success to decompose the polysaccharide by using enzymes derived from animal tissues. They further stated that it had not been possible to alter "the specific function [of the polysaccharides] by the action of any molds grown in solutions of the active materials. Experiments with molds, yeasts, and bacteria will be continued." [3] Although such experiments also failed, it is obvious that my own studies were only a continuation and extension of this initial program. The problem was still very much in Avery's mind when he first talked to me in 1927, and it was his vision of the potentialities of the enzymatic approach that set me on the way. Furthermore, it was his teaching skill that enabled me to assimilate rapidly the scientific lore and the techniques of his department, so that I could apply my specialized knowledge to the large problem he had formulated years before he began to indoctrinate me into biomedical research.

Such indoctrination of young scientists has been described by the late Colin MacLeod, who joined the department as a junior member of the clinical staff in 1935, and who stayed long enough to be one of the co-authors of the great DNA paper in 1944.

> For a time, the recent arrival saw little of the Professor. In some, this resulted in a sense of frustration at not being caught up immediately in the scientific life of the active department around them or being made a part of a current problem. Avery did not assign his associates to problems. His approach was indirect and at times seemed excruciatingly slow. After a week or two in the laboratory Avery commonly would invite the new assistant into his tiny personal laboratory. . . . A morning or afternoon would be spent in describing the lore of pneumococcus and in tracing the development of knowledge, the problems in which the department was currently concerned and those in which it had an interest. These soliloquies, prose masterpieces of high polish, were widely known as "Red Seal Records" [of which more later] and Avery was prone to repeat them as he sensed the necessity. If the candidate showed interest and began to read and work under his own steam, he was counselled and aided. A minimum of technical assistance was provided and one swam or sank because of one's own efforts or the lack of them. Avery placed emphasis solely on individual initiative and spurned team projects.[4]

As is clear from MacLeod's account, there was no organized teaching or training in the department; in fact, there was no formal organization of any sort. Avery never asked or urged anyone to do anything, to participate in any of his problems, or to initiate a new program. Consciously or unconsciously, however, he had developed a very effective technique to create unity of purpose among staff and visitors alike. The door of his office was always open, and he was ready at all hours of the day to welcome questions or statements from any one of us. In fact, except on very rare occasions, he acted as if he believed that the concerns we brought him were of major importance, but whatever the scientific problem discussed, he soon managed to emphasize one aspect of it that had a bearing on some phase of his own program. It did not matter whether the visitor's professional specialization was in clinical medicine, physiology, immunology, or chemistry, his attention was soon focused upon some aspect of the departmental problems to which his particular skill was well suited. That is the way Avery picked other people's brains, and also is the way he achieved unity of purpose within the department. The newcomer became a part of the team of his own volition almost unwittingly; he himself selected the area of work best suited to his own taste and gifts, while being gently maneuvered into one of the departmental problems.

This subtle manner of fostering cooperative action contributed greatly to the effectiveness and variety of the departmental research program. Its indirect consequences were even more important, for it gave each one of us the opportunity to discover our individual attributes and to gain self-confidence. Avery created an atmosphere in which our potentialities had a chance to emerge spontaneously. His department was a nursery in which any form of talent could unfold. One evidence that his teaching technique was effective is the high percentage of his collaborators who came to occupy important positions in medical schools or research institutes, and who continued to be productive investigators wherever they went. Few institutions can boast of such a large percentage of successful alumni!

The Protocol Experiment

As already mentioned, Avery was a late bloomer, but he moved fast after 1916. By 1923, he had become a full Member of The Rockefeller Institute, the highest rank in its scientific hierarchy. His fame was international when I became part of his department in 1927, but the physical and personal atmosphere in which he worked was still much the same as it had been when he joined the Institute in 1913. In fact, it remained essentially the same until he retired.

His laboratories were housed in a former hospital ward, still uselessly ornamented with quaint marble fireplaces. The high-ceilinged rooms were small and dissimilar in size; they were crowded with physicians and bacteriologists who were assisted by a few male technical helpers. Bacterial cultures were transferred and examined and serological reactions were carried out under conditions that would now be considered so primitive as to be incompatible with careful scientific work.

Most experiments were conducted at simple wooden desks that had been designed originally for office work and converted into laboratory benches by equipping them with microscopes instead of typewriters. The top of each desk accommodated a motley assortment of notebooks and simple laboratory instruments — test tube racks, glass Mason jars, droppers for various dyes and chemical reagents, tin cans holding pipettes and platinum loops — in brief, any object that might serve in bacteriological and serological manipulations. The same area was also used for handling experimental animals, for inoculating, bleeding, and dissecting them, and even for keeping some of them as pets.

The Bunsen burner on each desk served for aseptic transfer of cultures, heat sterilization, preparation of culture media, and also for some chemical operations. We used a great variety of kitchen utensils for many biological and chemical experiments. Needless to say, the laboratories dedicated to organic chemistry were equipped in a more sophisticated manner.

The larger pieces of equipment were situated in the middle of the rooms or wherever they could be fitted between the desks; they consisted mainly of a few simple incubators, vacuum pumps, and centrifuges. Each room had a single porcelain sink that served for almost any operation requiring the use of water, from staining slides for microscopic work to preparing extracts of bacterial cultures for immunological tests.

Avery's own laboratory was the smallest; it had formerly been the ward kitchen, and was equipped with the barest of bacteriological necessities. His office was adjacent to his laboratory and was, like it, small and bare. Both rooms were neat and clean, but kept as empty as possible, without the photographs, pictures, momentos, unused books, and other friendly items that usually adorn and clutter the working places of the white-collar class. The austerity of his office and laboratory symbolized how much he had given up in all aspects of his life for the sake of utter concentration on a few chosen goals.

His laboratory techniques were extremely simple, and he seemingly added to them only with hesitation and even reluctance. When the need arose, however, he went out of his way to learn new experimental procedures, as he did, for example, late in life during the phase of his work that led to the chemical isolation and identification of DNA; techniques were for him only a means to an end, and he never became a slave to them.

His kind of genius as an experimenter went far beyond that of the competent tradesman of science. First and foremost, it was marked by the thoughtfulness he applied to the selection of his distant goals, and the meticulousness with which he conducted his experiments. Then, when the results came in, he spent an incredible amount of time cogitating about their significance and distilling from them new interpretations of earlier knowledge and new ideas for further exploration.

Before deciding on any experiment, let alone starting it, he sat at his desk for days, mulling over the problem with associates or friends, and often alone. Among all the things that could be done, he was painfully anxious to determine by thought the one, or the very few, that appeared worth doing. He had developed an uncanny sense for recognizing what was truly important. Then, once he had made his choices, he formulated them over and over again, in as precise terms as possible. During that period, he consulted with anyone who was likely to contribute information relevant to the theoretical understanding of the problem at hand, or who was an expert in procedures that might serve in the experimental study. This was the time when he picked brains not only of his collaborators, but of anyone he managed to enlist under his flag by his persistence, his charm, and his enthusiasm.

Thinking, however, was never for him an end in itself. He had no taste for concepts that did not lead to experimentation. "Ideas are wonderful things," he would say, "but they don't work unless you work for them." Not only did he work for them; he put as much intensity into actual laboratory work as he did in thinking about experiments. The following statement by Maclyn McCarty is of interest because it is typical of Avery's attitude in the laboratory and because it refers to the period in the 1940s when he was working on the transformation of pneumococcal types after he had passed retirement age:

> Each morning on arrival in the laboratory the results of the experiment of the day before were waiting in the incubator to be read. Thus, when things were going well, each day began with a new bit of information that provided the stimulus and direction for further experiments. Fess and I had an unspoken agreement that prevented either of us from obtaining a sneak preview of the results before the other had arrived. . . . I recall the image of Fess as we converged on the incubator each morning, and in particular I see his expression, which was a curious mixture of eager anticipation and of apprehension for fear something had gone wrong with our complex biological test system — which, alas, was all too frequently the case.[5]

In view of this picture of Avery's childlike eagerness, it is rather surprising to read the statement by the late Arne Tiselius, as an explanation for the failure of the Nobel Committee to recognize Avery's DNA work, that "he was an old man when he made his discovery." [6]

Once experiments were under way, his obsession was to satisfy the most exacting criteria of evidence. He did not consider the work to be complete until all the results could be brought together in a perfect "protocol experiment" — one which incorporated all the variables and controls and which yielded the expected result without fail. The demonstration had to be so obvious that there was no need for statistical analysis. When this point had been reached, visitors and colleagues were invited to admire the simplicity of the experimental set-up and the clarity of its results. As long as Rufus Cole was director of the Hospital, he came down to the laboratory whenever the results were especially interesting, and joined in the chorus of admiration. We who had been involved in the excruciatingly slow, early phases of the work knew that endless discussions and numerous preliminary tests had preceded the experimental design that now appeared simple and decisive. For this reason, we were somewhat irritated to hear Dr. Cole tell us that we should learn from The Professor the art of planning and performing convincing experiments with small numbers of test tubes and animals, but nevertheless we enjoyed the show. The final demonstration with a few test tubes and a few animals never sacrificed any of the demands of scientific integrity; it was high-class showmanship, and had the quality of an artistic performance.

When an experiment failed to yield the expected results, extensive discussions of the new findings would ensue, with head-scratchings and exclamations of puzzlement. However, while the failure was the cause of much perplexity, it rarely led to a long period of discouragement. Avery's reaction was soon likely to be, "Now boys, whenever you fall, pick up something." And he would try to find in the unexpected result new ideas to be used for the problem at hand. In contrast to his cautious, conservative attitude while preparing an experiment, or during its performance, he would let his imagination be fired by any new challenge. During the initial period of puzzlement, he would formulate and encourage new hypotheses, some of them rather wild, in animated discussions and, even more often, in animated soliloquies. Almost any fact, however small and especially if unexpected, was likely to release in him a stream of theory.

Adrien Loir, Pasteur's nephew who assisted him in all his studies after his paralysis, has left a picture of the master's life in the laboratory that reminds me of Avery's behavior:

> Whenever there was a result, he [Pasteur] would build a whole new theory and expound it to anyone who was around; it is fortunate that there were few of us because it was a true novel [un véritable roman]. . . . He let his imagination run away when he was concerned with a particular topic. He would discuss his idea in the laboratory, at home, at the dinner table, everywhere. . . . He knew how to limit the numbers of his experiments, but in such a way that they gave an answer to his questions.[7]

When I read Loir's statement about Pasteur's "composition de ses romans," I remember Avery's mental constructs about the antigenic dissociation hypothesis that I shall discuss in Chapter Nine.

Even though he became readily intoxicated with his own ideas, Avery, like Pasteur, always retained his discipline as an experimenter. An orgy of talk was suddenly followed by phrases such as "We should be bold in formulating hypotheses, but we must be humble in the presence of facts," an expression which I believe he derived from Thomas Huxley. After the theoretical implications of new hypotheses had taken us into the stratosphere, he brought us back to a more sober view of reality with the homely reminder that the blowing of bubbles is all right as long as one remembers to prick them oneself. His eagerness to be the one to prick his own bubbles remained to the end a dominant aspect of his scientific attitude, as seen in the letter that he wrote to his brother Roy in 1943 to inform him of the role of DNA in the transformation of pneumococcal types (see Appendix I).

The Written Word

The meticulousness that Avery brought to the design, execution, and interpretation of experiments applied equally to the writing of scientific papers and the preparation of his rare public addresses. The process of organization, the balancing of one word or sentence against another, the discarding of draft after draft until the final product satisfied both his critical mind and his esthetic sense, brought him at times to a state bordering on neurosis. He ruthlessly destroyed all his preliminary texts, but fortunately Maclyn McCarty succeeded in salvaging one of the pencil drafts of the famous 1944 DNA paper. A page of it is illustrated in Figure 20.

He gave very few public lectures after he joined The Rockefeller Institute, but each was a matchless performance prepared with infinite care. He read the carefully written text with intense conviction and with a force that was more compelling for coming from such a frail body. All inflections of voice were tried repeatedly beforehand, using the laboratory staff and even casual visitors as sounding boards. Points of emphasis were indicated on the manuscript of the address (see Appendix II). The marvel of it was that, at the moment of the public performance, the speech was delivered in a quite natural tone, seemingly spontaneous, to such an extent that many listeners believed that much of it was improvised in front of them.

To submit a manuscript to him for discussion or approval was to impose on him a task at which he worked as hard as had the author. He analyzed the text for the quality of the scientific evidence and by identifying himself with the potential readers. In the words of Rollin D. Hotchkiss:

One's eyes were likely to be opened to undreamed of ambiguities and pitfalls that nest in the everyday language. A device he often used was to read aloud the prepared text, in the most gracious tones, but slyly emphasizing the wrong words, or pausing at the wrong places, so that new linkages were created, hanging participles were absurdly exposed, independent thoughts became comically interdependent, and the writer learned from a subtle master actor how weak the connection between thought and words can be.[8]

Although Avery worked so hard on his collaborators' manuscripts, he rarely, if ever, allowed his name to be listed as co-author unless he had participated in the experiments with his own hands. Quoting Hotchkiss once more:

I had always felt so deeply that I was an associate of Avery, that when preparing this article it was with great astonishment that I realized for perhaps the first time that we had never published a joint paper. The same association must have been felt by Drs. Frank L. Horsfall and George K. Hirst, to mention two virologists among many microbiologists who learned from him. Does the historian of science who leans heavily upon the printed word always learn of these vital but undocumented family pedigrees?[9]

After reading Hotchkiss's statement, I reviewed my own bibliography, and discovered that Avery's name appears on only four of the many papers that I published during the 14 years I worked in his department; these four all deal with the effects of the bacterial enzyme that decomposes the type III pneumococcal polysaccharide – a problem that he initiated and to which contributed directly, as I have reported. Yet he worked on all my manuscripts, including two that he tactfully put to rest in one of his drawers because he did not consider them worth publishing.

Avery derived much pleasure from the speculations that preceded and accompanied the writing of scientific papers, but he disciplined himself to indulge in such intellectual free-wheeling among only a few of his colleagues and friends, in unrecorded conversations. He saw no value in informing nonspecialists of all the preliminary stages in the establishment of facts or in the development of ideas. Just as he felt that scientific history is not illuminated by details of personal life, so he believed that the reporting of science is not served by the description of the uncertain steps that may or may not lead to worthwhile knowledge. He hated scientific gossip, and never repeated what he heard; nor did he ever make an unkind or invidious remark about any of his colleagues, even when he knew that they were critical of his work. Being acutely aware of human fallibility, including his own, he never engaged in public criticism or controversies, and he quietly ignored that which he could not believe.

Because "I" and "they" were words that had no place in his scientific vocabulary and because he labored endlessly on his papers to polish them and remove ambiguities, his style, though luminous, was rather impersonal. But this was the way he wanted it to be—an almost anonymous statement. By that very impersonal character, his style achieved a classical and austere quality that was a true expression of the way he controlled his nature and managed his life.

The Red Seal Records

When the United States entered World War I in 1917, Avery sought to obtain a commission in the Medical Corps of the U. S. Army, but this was denied to him because he had been born of British parents and was still a British subject. Even though he had lived in the United States for 30 years by the time the war broke out, he had never taken the trouble to become an American citizen! Eventually he enlisted as a private; because he was on active duty during hostilities, he qualified for immediate naturalization, and was commissioned captain.

He was fond of recounting his experiences of the period when, still a private, he had to lecture on bacteriology and infectious diseases to medical officers, many of whom held high military commissions. These officers were at first surprised and amused at the thought of being lectured to by such a small and unassuming teacher, but they soon recognized his technical competence and marveled at his authority and skill as a lecturer. He was dubbed once more "The Professor."

Although he retained the nickname "Professor" throughout his life, his lectures to Army officers during World War I were his last experiences with formal teaching. From then on, his influence was exerted almost exclusively in private conversations with the scientists who came to work in his department and with a host of people who were attracted by his reputation for knowledge and wisdom. Naturally, he had many visitors from other departments of the Institute and outside institutions, and from nonscientific circles. People came to him for advice on specific scientific problems and on personal matters, as well. I suspect, furthermore, that many came just to hear him talk and to watch him convert any situation or problem into an exciting display of words and gestures. He became a guru before the word entered American consciousness. I have already referred to him as a conversationalist, but I must come back to this aspect of his personality because it played an enormous role in shaping the scientific attitudes of his colleagues and, perhaps more importantly, in giving form to his own thoughts.

Although reluctant to speak at public meetings, coquettishly so, Avery was always eager to engage in conversations with colleagues, friends, or strangers. In general, however, the conversation soon evolved into a monologue, in the course of which his interests soon appeared more compelling and glamorous than the visitor's own concerns. These monologues had been thought out and were acted out in accordance with a carefully practiced formula. They were virtuoso performances, in which the theme was developed with logic and clarity, starting from the historical background and ending with the rationale of possible scientific approaches. The phraseology of these vignettes was remarkable. It included hesitations in speech, as if he were searching for a more accurate word or a more telling turn of phrase, whereas, in reality, the precision and effectiveness of the performance was the outcome of repeated rehearsals in the course of laboratory conversations. Several of us came to know various fragments of these conversation pieces by heart, and we referred to them as the Red Seal Records, after the name of the musical recordings that were then considered top-grade. In truth, there were times when we became somewhat impatient at hearing The Professor's Red Seal Records, but we were never bored by them, because we admired their artistic perfection. Furthermore, we realized unconsciously that they played an important role in the success of the department.

Avery's monologues certainly helped him to define his knowledge and to give structure to his thoughts. The continuous effort he made to sharpen and polish the language that he used to convey his concepts enabled him to recognize their ambiguities and inadequacies, and thereby facilitated the formulation of working hypotheses sufficiently well defined to be amenable to experimental testing.

My view of Avery's influence on the members of his department is based entirely on my personal experiences during the years of continuous and close association with him between 1927 and 1941. It is of interest to compare it with the views expressed by Rollin Hotchkiss, whose scientific background was very different from mine, and who worked with him between 1943 and 1948, during a period when the laboratory was no longer involved in problems of infectious diseases, but was concerned with genetic phenomena in pneumococci. Hotchkiss's impressions of Avery's attitude in the 1940s are so similar to what I remember of the 1920s and 1930s that I cannot refrain from quoting him at length at the risk of being repetitious. About the Red Seal Records, Hotchkiss writes:

> [Avery] successively played the parts of narrator, expositor, loyal opposition and finally attorney-in-summation. Even at the second or third hearing of one of these presentations, one could emerge, eyes glowing,

surprised to find that dusk had fallen outside while the new inner light was dawning.

These gems of perfection were continually revised and repolished. The highly organized presentation was a kind of debate with himself, punctuated with rhetorical questions like, "now, why should that be?" or "what does that all mean?" The auditor who was moved to try to respond, however, quickly found himself overwhelmed — and indeed suppressed — by the ongoing flow of well-rehearsed logic, that even in the voice of the man who seemed merely its spokesman, would brook no interference. These dissertations probably played a great part in concentrating the attention of his younger collaborators on basic problems, especially those involving that little gram-positive coccus which, he felt, presented in small compass most of the basic questions of biology.[10]

As we shall see, Avery did, in fact, touch on many crucial aspects of general biology in the Red Seal Records, even though his ideas were always based on the lore of the pneumococcus.

CHAPTER SIX

THE MULTIFACETED SPECIALIST

From 1913 to 1948, Avery occupied the same laboratory in the department of respiratory diseases of The Rockefeller Institute Hospital. Most of his experimental work was done with a single bacterial species: *Diplococcus pneumoniae*. The limited studies that he carried out with other biological systems were never far removed from pneumococcal biology. All his own experimental findings were published in the *Journal of Experimental Medicine;* the related aspects of his studies carried out by his chemist collaborators were published in the *Journal of Biological Chemistry*. All his public statements referred directly to his laboratory program, the only exception being his speech as President of the Society of American Bacteriologists, which was of a general nature, but which he did not allow to be published. It would be difficult, therefore, to imagine a more extreme case of scientific specialization. A rapid overview of Avery's scientific career will help to explain the paradox that, while he limited his investigations almost exclusively to pneumococcal disease, he managed nevertheless to throw light on a great diversity of theoretical problems in other fields of pathology and biology.

Avery is now chiefly remembered as a theoretical scientist, but this was only one aspect of his professional life. From the time he left medical school until the late 1930s, his dominant interest was the field of infectious diseases—how microorganisms invade the tissues and cause lesions; how the infected body responds to their presence; how recovery from infection takes place; how bacteriological and immunological knowledge can be used to develop rational methods of prevention and treatment.

Now that pneumonia can be treated readily with penicillin and other drugs, it is difficult to imagine what a distressing problem it was when the Hospital opened its doors in 1910. More than 50,000 persons died of the disease annually in the United States. It was more destructive than typhoid fever had ever been, and it replaced tuberculosis as "the captain of the men of death" among respiratory diseases. Physicians were so helpless against lobar pneumonia that William Osler referred to it as "a self-limited disease which can neither be aborted nor cut short by any means at our command." The patient either died irrespective of treatment or recovered spontaneously after approximately one week of acute disease — by "crisis" as the medical expression has been since Hippocrates. When Cole became director of the Hospital, he therefore decided that one of his major research goals would be to develop a therapeutic serum for pneumonia, and it was to this end that he appointed Avery as the bacteriologist on his team.

While working on the diagnosis and treatment of pneumonia, Avery became interested in the broader aspects of susceptibility and resistance to pneumococcal infections. He soon realized, however, that understanding such problems would require detailed knowledge of the pneumococcal cell itself, of its structure, its chemical composition, its physiological activities, its immunological characteristics, its genetic stability and variability. The systematic study of pneumococcal biology led him to deal with phenomena that transcended pneumococcal infections and that had, indeed, theoretical significance for unrelated biological problems. The following are a few of the discoveries he thus made:

— The virulence of pneumococci and of certain other bacterial species is conditioned by their ability to produce an ectoplasmic layer, which constitutes a cellular capsule. Encapsulated bacteria become avirulent when they lose the ability to produce the capsular substance.

— The capsule contributes to virulence by protecting the pneumococci against the defense mechanisms of the infected body, in particular against engulfment (phagocytosis) by the cells of the blood and tissues.

— Encapsulated pneumococci can be separated into types which differ chemically in the composition of their capsules. In all cases, the capsule is made up of a polysaccharide (complex sugar) but, because of chemical differences, each is immunologically specific for each pneumococcal type.

— The antibodies produced in the blood serum against the capsular polysaccharides protect against pneumococcal infection by neutralizing the antiphagocytic property of the capsules. The protection is specific for each pneumococcal type.

These facts, first established through the study of pneumococci, led to broader generalizations applicable to all infectious agents, for example:

— Minute differences in the chemical composition of microorganisms have profound effects on the response they elicit from animal and human tissues. As a consequence, specificity can be defined in the precise terms of molecular chemistry, not only in the case of polysaccharides, but also of other types of chemical substances.

—The ability of microorganisms to survive and multiply in the body depends on their possession of specialized cellular structures. Virulence, which used to be regarded as a mysterious attribute of certain microbial groups, can be explained by the interplay of these structures with the body's defense mechanisms.

These discoveries soon established Avery as one of the world's most original investigators in the field of infection and immunity. Then he progressively became involved in another line of work, which led to his most spectacular achievement, this time in the field of genetics. He demonstrated that bacteria can be made to undergo hereditary changes by treating them with deoxyribonucleic acids (DNA) extracted from other bacteria. This discovery turned out to be one of the landmarks of modern biology, because other investigators soon established that DNA molecules are the specific carriers of hereditary characteristics in *all* living things.

Remarkable as Avery's achievements were, they had no obvious interest for the general public; there was nothing in them comparable in excitement value to the discovery of a new drug or vaccine. Who but a theoretical scientist could be impressed by the fact that subtle differences in a sugar molecule condition the fate of microbes in the body? That certain virulent bacteria become innocuous when they lose the ability to produce an ectoplasmic layer or some other cellular constituent? That the chemical composition of the pneumococcal capsule is genetically determined by a specific deoxyribonucleic acid?

When Avery died, the writers of newspaper obituaries tried to glamorize his work by asserting that it had led to the development of miracle drugs, but this was not true, and he would have been deeply embarrassed by such a statement. In fact, his discoveries had only few and limited applications of immediate practical importance. His contributions were to the understanding of biological phenomena, and have influenced the practice of medicine only in an indirect manner. He advanced biological and medical sciences by providing factual knowledge and thought patterns applicable to the study of all living things; in particular, he demonstrated the possibility of a chemical approach to the understanding of infection and of heredity—hardly a topic to make newspaper headlines!

Avery's characteristic habit of intellectualizing a problem, then converting it into concrete chemical operations, is illustrated in the following technical chapters, which deal with the bulk of the experimental work he conducted at The Rockefeller Institute in the fields of immunity, virulence, and heredity. Most of his findings have now been incorporated into theoretical biology and medicine, but some have been questioned, or even shown to be erroneous. For example, he failed to establish his early hypothesis that resistance to infection with pneumococci results from the inhibition of pneumococcal enzymes by the blood serum of infected persons or animals. Although this metabolic approach to the immunity problem was abortive, it provides a useful introduction to his experimental work at the Institute, in part because it was his first original investigation in the field of pneumonia and, more interestingly, because it reveals how his propensity to construct mental pictures of biological phenomena influenced his vision of scientific reality.

CHAPTER SEVEN

THE LURE OF ANTIBLASTIC IMMUNITY AND THE CHEMISTRY OF THE HOST

Antiblastic Immunity

Avery was a persistent man. Once he became involved in a scientific problem he pursued it doggedly, waiting, if need be, for many years until he saw the way to a solution. He even pretended at times that he enjoyed the failures that are inevitable in scientific life. "Disappointment is my daily bread," he was wont to say. "I thrive on it." By this he probably meant that his eagerness to overcome difficulties generated in him new strength and new inspiration. However, there is at least one topic—antiblastic immunity—that he seemingly abandoned after having affirmed its importance in the annual reports to the Board of Scientific Directors of The Rockefeller Institute for 1915 and 1916, and more publicly in an article that he and Dochez submitted to the *Journal of Experimental Medicine* in July, 1915, and that was published in 1916.[1] He never again mentioned antiblastic immunity in print after this publication, but he kept it in mind to the end of his professional life.

In their *Journal of Experimental Medicine* paper, Dochez and Avery defined antiblastic immunity as a resistance to pneumococci, resulting from the inhibition of certain pneumococcal enzymes by the serum of the infected person or animal. The origin of the theory can be traced to experiments briefly mentioned in the report of April, 1915, in which Avery appears for the first time as an original investigator.[2]

When he joined the Hospital staff in 1913, he was responsible for the isolation of pneumococci from patients and also for the production and testing of antipneumococcal sera. In the course of his routine duties, he noticed that, although therapeutic sera do not kill pneumococci, they retard their growth in culture media; he gave reasons to support his belief that this growth-inhibitory effect was not due to agglutination of the organisms by the serum. Shortly after this initial observation, Dochez and

Avery found that antipneumococcal serum "inhibits certain digestive and fermentative properties of the pneumococcus, especially the formation of amino acid from protein, and the fermentation of various carbohydrates."[3] It is probable that Dochez was led to this enzymatic explanation because, a few years earlier, he had collaborated with Eugene Opie on the enzymes of inflammatory cells.

The facts observed by Dochez and Avery were few and limited in scope, but the two scientists did not hesitate to elaborate from them a bold metabolic concept of immunity. They based their argument on the hypothesis that pneumococci could not multiply *in vivo* if they were not capable of utilizing certain constituents of the infected host by means of enzymes located at their cellular surface. They suggested, furthermore, that any mechanism interfering with the action of these surface enzymes would provide resistance to pneumococcal infection. In their words, "We have chosen the term 'antiblastic immunity' . . . to indicate that the forces at work are antagonistic to the growth activities of the organism." (The word antiblastic is derived from *blastos,* the Greek word for sprouting, or growth.[4]) This definition made it clear that the kind of resistance to infection they had in mind was completely different in mechanism from that induced by the conventional protective antibodies.

The experiments described in the 1916 paper unquestionably show that addition of antipneumococcal serum to a broth culture of pneumococci retards the growth of these organisms and partially inhibits some of their enzymatic activities. They also establish that this so-called antiblastic property appears in the serum of patients suffering from lobar pneumonia at the time of the crisis, more or less concurrently with the appearance of the type-specific antibodies usually associated with recovery from the disease.

As judged from the annual reports, Dochez and Avery continued to work on antiblastic immunity for at least several months after their paper had been submitted to the *Journal of Experimental Medicine.* In the report for 1916, they described the new finding that the metabolic functions of pneumococci "recently isolated from the human body are more resistant to the inhibiting factors of both normal and immune sera than are those of organisms which have led a saprophytic existence for considerable periods of time."[5] The obvious implication of this statement is that the virulence of a microorganism is a consequence of its ability to resist the antiblastic mechanisms of the human body.

In the 1916 *Journal of Experimental Medicine* paper, Dochez and Avery stated that if the theory of antiblastic immunity turned out to be true, "considerable light would be thrown on the obscure mechanisms by which parasitic bacteria establish themselves in animal tissues, and on the forces mobilized by the animal body in opposition to such invasion."[6] They also formulated the more adventurous hypothesis that in pneumococci "capsule formation represents on the part of the organism an attempt to protect the function of its digestive zone." In the annual report for 1916, they made the further unorthodox statement that "for the animal body to rid itself of infection, the growth of the infecting microorganism *must first be arrested* and that *only after this has occurred* do the more specific substances have an opportunity to exert their full effect" (italics mine).[7] This was a revolutionary view of resistance to infectious disease, since it implied that the hypothetical mechanisms involved in antiblastic immunity had to stop the growth of the infectious agent by inhibiting its enzymes before antibodies and other immunological mechanisms of resistance could come into play.

Ironically enough, Dochez and Avery themselves were soon to direct the study of resistance into immunological channels by their 1917 discovery of the specific soluble substances of pneumococci (see Chapter Eight). Until 1916, however, they apparently believed that antiblastic immunity was at least as important as conventional antibodies in providing resistance to pneumococcal infection and in recovery from lobar pneumonia.

Some remarks made by Dr. Rufus Cole in his annual report for 1919 suggest that he had doubts concerning the significance of what he referred to as "the question of *so-called* antiblastic immunity" (italics mine).[8] In 1917, Francis Blake, who was himself at The Rockefeller Institute Hospital, published a paper denying that there was any such thing as antiblastic immunity.[9] He had confirmed Dochez's and Avery's findings that, under the conditions they used, antipneumococcal serum can indeed retard the growth of pneumococci and depress their enzymatic activity, but he asserted that these effects were simply the result of agglutination of the organisms, and did not involve inhibition of their enzymes. Equivocal results were also published in 1919 by M. A. Barber, another bacteriologist working at the Hospital.[10]

Neither Dochez nor Avery published anything to refute their colleagues' criticisms. In fact, they never mentioned antiblastic immunity again in either scientific journals or the annual reports. One could therefore conclude that they rapidly lost interest in the subject and dismissed it from their minds, but this is not the case.

On several occasions during the 1930s, both Avery and Dochez discussed with me the possible role of antiblastic immunity in pneumococcal infections. Their reference to this topic may have been in part an expression of their usual graciousness, since they knew that the subject was close to my own scientific interests and was, indeed, directly relevant to my experimental work. However, I have been told by Dr. Maclyn McCarty that he heard Avery mention antiblastic immunity during the 1940s, at a time when the dominant concern of the department was the identification of the substance responsible for the transformation of pneumococcal types. There are reasons to believe that much of the experimental work in Avery's department for almost 20 years derived from his early hunch that the metabolic activities of pneumococci are an essential aspect of their role in infection.

Before proceeding to that topic, however, I shall open a parenthesis to express my personal view that the problem of antiblastic immunity should be reinvestigated. Admittedly, the experiments published by Dochez and Avery in their 1916 paper do not prove the validity of the hypothesis, but neither do the papers published by Blake in 1917 and Barber in 1919 rule the phenomenon out of existence.

During the past few decades, several experimental systems have been studied in which control of experimental infection is brought about by inhibition of the pathogen's growth, not by its destruction, and this is precisely how antiblastic immunity is defined. Experimental infections caused by certain protozoa and by tubercle bacilli are cases in point. Furthermore, all enzymes that have been studied from the immunological point of view have been found to be antigenic, and the antibodies that they elicit have some inhibitory effect on their enzymatic activity.[11] It is of interest, in this regard, that such inactivation does not take place if the enzyme is intracellular.[12] This gives added interest to the suggestion by Dochez and Avery that the antiblastic effect takes place "at the surface of the bacterial cell," and that the integrity of this zone "is essential to the growth of the bacterium." While they had no evidence for their hypothesis, their words acquire a prophetic quality when read in the light of present information about the importance of the cell membrane in nutritional and developmental processes. The plasma membrane, as is well known, is one of the important organelles of the cellular structure. Thus, although antiblastic immunity was a hypothetical concept half a century ago, it can now be reformulated in more precise terms and put to experimental tests.

Whatever its possible scientific importance for the understanding of infectious processes, the subject of antiblastic immunity may have had a decisive role in Avery's psychological development; it probably accounts in part for his fear of making public statements that went beyond well-established facts.

The hypothesis that he and Dochez formulated in their *Journal of Experimental Medicine* paper and in the annual reports for 1915 and 1916 gives the impression that the two eager scientists had concocted the antiblastic theory on the basis of abstract thought in the course of their midnight discussions. As it turned out, the few laboratory tests they conducted were sufficiently encouraging to give substance to the products of their imagination; therefore, they felt justified to compound hypothesis with hypothesis into a sweeping metabolic theory of virulence and immunity. As we have seen, however, the interpretation of their findings was questioned by some of their own colleagues, and they themselves soon discovered new facts that pointed to a mechanism of resistance to pneumococci in which enzyme inhibition had no part.

Avery again published statements for which he had inadequate evidence, and which were soon proved to be wrong, when he stated in 1917 that the specific soluble substances of pneumococci are of protein nature and are responsible for the toxic effects of pneumococcal infections (Chapter Eight). But after these mistakes, he learned his lesson. Never again did he mention in scientific journals the products of his imagination, unless they had been documented by overwhelming laboratory evidence. He continued to indulge in fanciful scientific theories throughout his professional life, but only in the course of private conversations among colleagues and a few friends, or now and then in the annual reports of The Institute. It is probable that his experience of 1916 with antiblastic immunity, then of 1917 with the specific soluble substance, contributed to his conservative behavior in all his public statements and to his eagerness that he should be the one to prick his own bubbles.

Bacterial Metabolism and the Phenomena of Infection

Two high points in Avery's scientific life occurred in 1917, when he and Dochez first reported on the specific soluble substances of pneumococci, and in 1923, when he and Heidelberger demonstrated to everybody's surprise that the immunological specificity of these substances was to be found in their polysaccharide nature. One might assume that immunochemistry had been his dominant scientific concern between these two dates. In reality, more than 30 papers dealing with pneumococcal enzymes, accessory nutritional factors, and other topics of bacterial metabolism were published from his laboratory in the six years between 1919 and 1925; his name appears as co-author in 23 of these studies, and it is certain that he participated very directly in planning all the others and writing their results.

The introductory statements of Avery's papers on bacterial metabolism and nutrition, as well as the discussions and summaries that conclude them, are written as if the motivating force behind this impressive amount of experimentation was his interest in the characteristics of pneumococci considered as independent organisms. A different interpretation emerges, however, from his statements in the annual reports to the Board of Scientific Directors.

The section dealing with antiblastic immunity in the annual report for 1916 emphasizes the difficulties that Dochez and Avery experienced when they tried to quantitate the inhibitory effect of serum on the enzymatic activities of pneumococci. The major source of these difficulties was that, under the conditions of their tests, the enzymes themselves underwent spontaneous inactivation because of the rapidity with which living pneumococci disintegrate and autolyse in artificial media. In order to deal with this problem, Avery decided to separate the enzymes from the bacterial cells, on the assumption that he could obtain them in a stable form. He was sufficiently successful to express the view in the 1916 report that "much more reliable" results could be expected in the future because he had obtained from pneumococci a "proteolytic enzyme" that was "active in the absence of the living cell."[13] This statement reveals an attitude toward biological work that he was to maintain in all his subsequent investigations — that the way to study a biological phenomenon is to obtain the substance responsible for it in an active, stable form, and to determine its chemical activities under controlled conditions. From a more limited and immediate point of view, his attempts to measure antiblastic immunity by chemical methods, as described in the 1916 report, heralded a long program of biochemical studies encompassing topics ranging from metabolic equipment of pneumococci to bacterial nutrition to oxidation-reduction processes.

From 1918 to 1925, Avery devoted a very large percentage of his time to biochemical studies of bacteria in collaboration with G. E. Cullen, T. Thyotta, H. J. Morgan, and J. M. Neill. When I joined his department in 1927, he encouraged me to develop this program on my own. He himself could no longer participate directly in its execution because of his involvement in immunochemical studies, but he was intensely interested in its progress, and gave extensive coverage to it in his annual reports.

Much of the departmental work on bacterial nutrition and metabolism was focused on problems of immediate practical importance in laboratory experiments. In particular, it aimed at improving culture media either for diagnostic purposes or for producing large quantities of bacteria, and later at assuring the genetic stability of bacterial strains. Some of the findings, however, had a larger biological significance. Avery himself worked intensely on the recognition in bacteriological culture media of two accessory growth factors that he designated as the X and V factors, and that were later identified in other laboratories as heme and cozymase. Commonplace as this knowledge is today, in 1921 the recognition of the X and V factors was an important step in the development of the science of bacterial nutrition.

There was always in the background, furthermore, the hope that studies of bacterial metabolism would eventually contribute to the understanding of pneumococcal disease — of natural resistance to it, of recovery from it, of its pathogenesis and its epidemiology. This hope can be illustrated by many statements taken from the annual reports.

In the April section of the 1920 report, a series of papers from Avery's group, published in the *Journal of Experimental Medicine* under the general heading "Studies on the enzymes of pneumococcus," was justified as follows: "With the hope of acquiring a more definite understanding of the way in which pneumococci adapt themselves to various environments, a study is being made of the enzymes of pneumococcus."[14] In the October part of that report, Avery expressed the belief that "these studies are important, not only theoretically because of the added knowledge gained concerning life processes of the organism, but also clinically."[15] In the 1923–24 report, he again justified his interest in bacterial metabolism with the statement that enzymatic and metabolic studies had "yielded certain facts which are not only of interest with reference to the physiology and chemistry of the bacterial cell, but which give promise of wider significance in the interpretation of the process of infection in the animal body."[16] He was aware that experiments carried out in other institutions had shown that the activity of certain enzymes can be inhibited by antibodies prepared against them, as he and Dochez had postulated in their 1916 paper.

The Chemistry of the Host

A favorite subject of laboratory discussion in the 1930s, and one commonly mentioned in the annual reports, was that even the most virulent pneumococci are incapable of initiating infection until the normal defensive mechanism which guards the lower respiratory tract has been broken down. There was no understanding of the factors involved in this breakdown except that the general physiological condition of the host was involved. It was known, for example, that the incidence of lobar pneumonia was in some way correlated with attendance at football games, and probably with the excessive consumption of liquor on such occasions. Many experiments were performed with a variety of animals in an attempt to overcome the normal resistance of their lungs to pneumococci. For this reason, mice intoxicated with alcohol and exposed to sprays of pneumococci were a common sight in E. G. Stillman's laboratory across the hall from Avery's office. Although the results of this type of experimentation were meager, they provided food for much thinking and guessing about the interplay between the physiological characteristics of pneumococci and the chemical conditions in infected organs.

From 1925 on, the pressure of the immunochemical studies with capsular polysaccharides and with synthetic antigens, and the phenomenal success of these studies, prevented Avery from developing further his interest in the metabolic and physiological aspects of infection, but he came back to the subject through an accidental discovery.

While following the development of specific antibodies in the serum of pneumonia patients, he and T. J. Abernathy recognized the occurrence of a peculiar serum protein not normally present in the blood. This substance did not behave like an antibody, because it appeared very early during the acute phase of the disease and disappeared within 24 hours after recovery. The newly discovered serum factor was dubbed the "C-reactive protein" in laboratory parlance, because it precipitated in contact with the so-called "C polysaccharide" of pneumococci. Avery saw in this unexpected observation an opening into aspects of the infectious process that transcended the classical antigen-antibody reactions, and that might eventually throw light on the physiological responses of the body to pathological phenomena. An indication of his interest in this new aspect of host-parasite relationship appears in a small item reported by Rollin Hotchkiss: "In 1938 . . . I begged for an opportunity to work on transformation but he [Avery] was anxious to further the work on blood proteins in acute infections and asked me to wait."[17] Avery certainly welcomed the prospect of being able to anchor the physiological problem of host-parasite relationship on a substance that could be isolated and chemically characterized.

Although he and Abernathy had first detected the new protein in the serum of pneumonia patients, they soon found that it also occurred in the serum of patients suffering from many acute infections caused by other pathogens. Three papers published in 1941 defined the characteristics of the C-reactive protein, and established that it was not an antibody. They provided evidence, instead, that it was released from tissues in the course of infection, probably as a result of some cellular damage, and that it therefore represented a nonspecific reaction of the tissues to injury. The C-reactive protein was isolated as a highly purified, immunologically homogeneous protein by Avery and MacLeod in 1941, and crystallized shortly after by McCarty.

Avery never discussed the C-reactive protein without turning the conversation to what he was wont to call "the chemistry of the host." Although he never spelled out what he meant by that expression, he clearly had in mind all the unidentified body substances and mechanisms of a nonimmunological nature, both protective and destructive, that come into play in the course of infectious processes. Along with the C-reactive protein, he probably would have listed in this category other products of cellular damage or stimulation, the multiple aspects of the inflammatory responses, and—even though he rarely mentioned them—the mechanisms responsible for antiblastic immunity.

Host-Parasite Relationships

The intensity of Avery's interest in the physiological determinants and products of host-parasite relationships comes to light in a written text on which he worked for many days but refused to publish—the speech he delivered in 1941 as President of the Society of American Bacteriologists (Appendix II).[18] In it, he quoted at length from the prophetic words of the seventeenth-century English chemist Robert Boyle, who guessed as far back as 1660, two centuries before Pasteur, that the mechanisms of disease would eventually be explained through an understanding of fermentations. In Boyle's words:

> He that thoroughly understands the nature of ferments and fermentations shall probably be much better able than he, that ignores them, to give a fair account of divers phaenomena of several diseases (as well fevers as others) which will, perhaps, be never thoroughly understood, without an insight into the doctrine of fermentation.[19]

After quoting Robert Boyle, Avery went on to discuss the wide range of interplay between chemistry and microbiology; this gave him the opportunity to emphasize the solidarity of the different fields of natural science and the dangers of scientific specialization. I shall come back later to the general aspects of this speech (Chapter Thirteen).

Contrary to what could have been expected from a scientist who had devoted most of his professional life to the study of specific immune mechanisms, Avery did not mention antigen-antibody reactions or cellular immunity. Instead, he forcefully stated that the best chance for progress in the control of microbial diseases "lies not alone in the discovery of ways and means of fortifying the natural and specific defenses of the host—

important as these are—but in a better insight into the *structural and cellular mechanisms of host and parasite*" (italics mine).[20]

He concluded this section of his speech by paraphrasing Robert Boyle to predict that the phenomena of infectious diseases will never be "thoroughly understood without an insight into the life processes of the host and parasite."[21] This statement calls to mind similar ones that he and Dochez made in their 1916 paper on antiblastic immunity, when they tried to explain in metabolic terms the phenomena of bacterial virulence and host resistance.

The members of the Society of American Bacteriologists who listened to Avery's speech in 1941 knew, of course, that his most spectacular achievements at that time had been in the field of immunochemistry; many of them also were familiar with the work of his department on bacterial transformation. They were probably surprised and, perhaps, disappointed that he did not refer to any of those fundamental studies, but chose instead to discuss in rather vague terms the cellular structures and metabolic activities of microorganisms and the chemistry of the host. Yet, what they heard was far more important than what they had expected, and also a far more revealing expression of Avery's genius as a scientist. He could have discussed conventional knowledge and theories, but he pointed instead to problems that had not yet been defined; he could have reminisced about the past, but he looked into the future and suggested work to be done.

In expressing his faith that studying the interplay between the life processes of the host and those of the parasite was the way of the future, he was continuing the midnight discussions he had begun 30 years earlier with Dochez. He was still cogitating about the precise nature of the physiological processes that determine the outcome of host-parasite relationships and that he had symbolized by the expressions antiblastic immunity and chemistry of the host.

CHAPTER EIGHT

THE CHEMICAL BASIS OF BIOLOGICAL SPECIFICITY

Serum for Pneumonia

As mentioned in Chapter Six, one of the chief projects at the Rockefeller Hospital was the development of a therapeutic serum for lobar pneumonia. The problem appeared well defined in 1909, when Cole first formulated his research program. Most cases of lobar pneumonia were known to be caused by the pneumococcus, a delicate microbe that had been described in 1881 by Pasteur in France and almost simultaneously by Sternberg in New York—the same Sternberg who was, for a while, director of the Hoagland Laboratory where Avery began his scientific life. The causative role of the pneumococcus in human pneumonia had been demonstrated between 1884 and 1886 in Germany. Good techniques were available to isolate the microbe from patients, to cultivate it *in vitro*, and to produce with it an experimental disease in animals. The problem selected by Cole thus appeared straightforward: to immunize horses against pneumococci and to administer the antipneumococcal serum thus obtained to patients suffering from pneumonia.

An unexpected difficulty arose just as the work was being planned. In 1909 and 1910, Neufeld and Handel, of the Robert Koch Institute in Berlin, reported that they had found subtle, but important, differences among the various cultures of pneumococcus isolated from patients. What had been regarded as the pneumococcus species turned out to consist of a variety of strains, which, although similar in appearance and in general characteristics, differed in immunological properties. The German workers divided the pneumococcus group into three well-defined types and a heterogeneous subgroup. These findings greatly complicated the problem of serum therapy, because the serum prepared against any one particular pneumococcal type was inactive against the other types.

Dochez was given the task of comparing the distribution of pneumococcal types in New York with that found by Neufeld and Handel. In 1913, he reported that the cultures he had studied could be divided into three main types identical with those found by the German workers, plus a fourth group, made up of poorly characterized subtypes; this fourth group was dubbed by some English bacteriologists, somewhat contemptuously, as the American Scrap Heap.

Therapeutic trials conducted in the Hospital with antipneumococcal sera soon gave encouraging results in type I lobar pneumonia. As serum was not commercially available, a program of production was developed at the Institute. Avery was given the responsibility for the vaccination of horses, the processing of serum, and the measurement of its antipneumococcal activity. He was also made responsible for much of the diagnostic work, and developed a rapid culture method for determining the pneumococcal types recovered from patients. The mastery and authority he acquired within a few years can be measured from the fact that in 1917, less than four years after joining the Hospital, he was the senior author of a classic monograph entitled *Acute Lobar Pneumonia: Prevention and Serum Treatment,* published by The Institute.[1] In this monograph, he, in collaboration with H. T. Chickering, Dochez, and Cole, set forth everything they had learned about lobar pneumonia from their practical experience in the laboratory and on the wards: the relative prevalence of the various pneumococcal types; how to prepare an effective serum against type I; and how to treat the disease.

Avery's preoccupations with pathological processes were at the basis of the biochemical investigations mentioned in the preceding chapter and of the immunological investigations that will be considered. Much of his work was conditioned by his desire to produce effective therapeutic sera. Until the advent of chemotherapy, this was the rallying point in his department; it kept everybody's eyes on the ball. Many abstruse studies on "antigenic dissociation" and on the autolytic system of pneumococci, which will be discussed later, appear far removed from the clinical problems of disease, but they were, in reality, steps toward a better understanding of the factors involved in immunity to pneumococcal infection and in the production of therapeutic sera.

In the 1936–37 annual report to the Corporation, Dr. Cole detailed some of the theoretical factors that had led the Avery department to shift from horses to rabbits for the production of antipneumococcal sera. The technical aspects of this problem cannot be discussed here, but the clinical results deserve to be mentioned, because they demonstrate that Avery's continued interest in serum production had important practical consequences. Cole reported:

Among more than fifty patients suffering with lobar pneumonias due to pneumococcus Types I, II, V, VI, VII, XIV, XVIII there has been but one death and this occurred in a patient five weeks convalescent from pneumonia. In untreated patients with similar type distribution the death rate would have amounted to about 34 percent. . . . In the last several cases treated with immune rabbit serum in this hospital, the average time from the first injection of serum until the crisis was less than five hours. In many patients normal temperature, pulse and respiration were regained in as short a time as five hours after serum was administered.[2]

Admittedly, the therapeutic results were not satisfactory in pneumonia caused by type III pneumococci; furthermore, the need for rapid typing of the pneumococcus cultures in each individual patient and other practical problems still stood in the way of the widespread use of antipneumococcal sera in general practice. Admittedly, also, the introduction of sulfa pyridine and other sulfa drugs in the late 1930s, then of penicillin in the early 1940s, made serum therapy obsolete within a very short time after it had been perfected. This, however, does not detract from the scientific and practical quality of the research program that led to the development of antipneumococcal sera; in fact, the achievement remains one of the finest examples of the application of orderly, rational thought to a therapeutic problem.

The Specific Soluble Substances

During late 1916, Dochez discovered in the filtrate of a pneumococcus culture a soluble substance that flocculated in the antiserum prepared against the particular type of pneumococcus growing in the culture. Further tests soon revealed that this was a general phenomenon. Each of the various types of pneumococci isolated from patients was found to produce in culture media a soluble substance that possessed its own type specificity. Avery soon joined forces with Dochez in the analysis of the phenomenon, and together they established that the specific soluble substances, designated SSS in laboratory parlance, were released in solution early during the life of the pneumococci, and therefore were not the products of bacterial disintegration.

Dochez and Avery also established by immunological techniques that the SSS passes into the blood and urine of patients during the acute phase of pneumococcal pneumonia. Both were fond of telling how, after having first detected the substance in the blood, they reasoned that, since it was soluble and diffusible, it might filter through the kidneys into the urine. They accordingly requested from the ward a urine specimen from a patient with a severe type II pneumonia and tested it with type II antipneumococcal serum. To their great disappointment, no flocculation occurred. They sat glumly looking at the tubes, which had remained clear, wondering what was wrong with their reasoning. After a while, Avery walked to the vase of urine received from the ward, picked it up, and looked at the label; to their relief, it turned out that the specimen was from the wrong patient. The precipitin test was positive when carried out with the proper specimen. SSS did indeed pass through the kidneys into the urine.

The discovery of SSS, and the demonstration of its presence in the body fluids of patients, were published in 1917 by Dochez and Avery in two papers that are now classic.[3] As Dochez was away in France during the early part of 1917, Avery was left alone to continue the work. He obtained further evidence that SSS can usually be found in the urine of pneumonia patients, and he made use of this observation to develop a diagnostic test that in many cases permitted rapid identification of the pneumococcal type, even before the culture was recovered from the patient.

The annual report for April, 1917, shows that he had already been bitten by the desire to know the chemical nature of the specific soluble substances. He separated serologically active material from the urine "by repeated precipitation with acetone and alcohol," [4] and the simple chemical analyses that he carried out on it led him to state in the report: "The determination of total nitrogen and nitrogen partitions . . . shows that this substance is of *protein nature* or is associated with protein" (italics mine). This statement, which he himself, with Heidelberger, later proved was erroneous, found its way onto page 493 of "The elaboration of specific soluble substances by pneumococcus during growth," a paper he published with Dochez in the *Journal of Experimental Medicine* in 1917.[5]

The first World War naturally disrupted the research program at The Rockefeller Institute. Until January, 1919, Avery ran a course for Army medical officers on the "etiologic diagnosis of acute respiratory diseases." The epidemic of so-called Spanish influenza compelled The Rockefeller Institute Hospital group, under the leadership of Rufus Cole, to become involved in studies of influenza bacilli and hemolytic streptococci, organisms that were frequently found in the pneumonia outbreaks studied first in Army camps, then in the civilian population. Avery published papers on the bacteriology of these two species in 1918 and 1919. Then, immediately after the war, he turned his attention to the metabolic studies that have been discussed in the preceding chapter.

Six years elapsed between the original publication on the specific soluble substances in 1917 and the paper on the chemical nature of these materials that Avery published with Heidelberger in 1923. He had not been idle during that period; he published 14 papers on several entirely different topics, but none of these dealt with immunological problems. One can surmise from indirect evidence, however, that he had become convinced in the meantime that the specific soluble substances are the main, if not the only, components of the capsules that surround the cells of virulent pneumococci. It is certain that he continued to work on the purification of the capsular substances by his simple techniques of kitchen chemistry, and it is probable that he came to question the validity of his 1917 statement that they were "of protein nature." There is no doubt that he was asking himself the kind of question that henceforth would remain the leitmotiv of his research. When studying a biological phenomenon, he would always wonder, "What is the substance responsible?" and "How does it work?" As I have recounted in a preceding chapter, this is the kind of question he had in mind when, in 1922, he finally managed to secure the collaboration of Michael Heidelberger.

The rapid success of the Avery-Heidelberger collaboration is demonstrated by the fundamental papers on immunochemistry that they published together between 1923 and 1929 (Heidelberger left The Rockefeller Institute in 1927 for The Mount Sinai Hospital in New York). These papers are readily available, but in discussing the experimental work on which they are based, I shall quote chiefly from the annual reports, which give a clearer impression of Avery's hopes and of his constant worries about the significance of his work.

From the beginning, he was interested not only in the chemical nature of the capsular substances, but also in the general problem of biological specificity, a concern clearly formulated in the annual report for 1922–23. In it, he recalled his 1917 investigations, which had established "that the specific substance . . . is precipitable in acetone, alcohol and ether; that it is precipitated by colloidal iron and does not dialyze through parchment; that the serological reactions of the substance are not affected by proteolytic action by trypsin." [6] He saw in these properties *"an ideal basis for the beginning of a study of the relation between bacterial specificity and chemical constitution"* (italics mine).[7] He expressed the same view in the annual report for 1923–24, in which he stated that the specific soluble substance "was selected as a basis for the present studies on the chemistry of bacterial specificity because it was not only highly type-specific, but also possessed a stability to heat, enzymes, and many chemical reagents that augured well for its suceptibility *to study by the methods of organic chemistry"* (italics mine).[8] This phrase makes clear that his goal was not merely the chemical isolation of the substance, but the establishment of the molecular basis of immunological specificity.

Avery and Heidelberger started their work with the type II specific soluble substance, and soon established that the preparations possessing specific immunological activity consisted predominantly of complex polysaccharides. They felt, however, that this was not sufficient evidence to prove the chemical nature of the active material, because, in their words, "it seemed possible that the polysaccharide . . . might be a tenaciously adhering impurity, and that the actual specific substance might belong to some other class of organic substance."[9] After submitting the active material to a variety of purification procedures and chemical analyses, they finally concluded "that the low nitrogen content, 0.1–0.2 percent, and absence of reactions for protein split products exclude relationships with . . . proteins and their derivatives." They noted furthermore "that by all the methods hitherto used for purification . . . essentially the same type of polysaccharide is recovered. It is, therefore, becoming increasingly difficult to believe that the carbohydrate present can be merely a tenaciously adhering impurity."[10]

I have quoted these passages to illustrate that Avery was acutely conscious of the possibility that the immunological specificity of the soluble substance was due to a contaminating protein; the phrase "tenaciously adhering impurity" occurs twice in the report.

The announcement in 1923 that the immunological specificity of type II pneumococcus is due to a polysaccharide was greeted with wide skepticism and even sarcasm. It went counter to the orthodox view that only proteins are sufficiently complex in structure to allow for the enormous degree of diversity required to account for immunological specificity. But Avery and Heidelberger refused to become involved in controversies; they eventually convinced their critics by the sheer accumulation of new facts.

Within a year, work with type III specific soluble substance "showed that marked chemical differences existed between it and the corresponding substance of type II." The type III substance appeared to be "an optically levorotatory strong acid, hydrolyzing to reducing sugars, chief of which is perhaps glucuronic acid or an analog, and not glucose, as in type II."[11] Thus was laid the groundwork for all the immunological studies which led, a decade later, to the synthesis of artificial antigens.

In 1924, Avery and Heidelberger were joined by the organic chemist Walther Goebel, who determined with greater precision the molecular structure of the various pneumococcal capsular polysaccharides. Another bacterial species, *Klebsiella pneumoniae* (designated Friedländer bacillus in the Avery publications), was added to the immunochemical program because the virulent forms of this organism are occasional causes of lobar pneumonia and are encapsulated as are pneumococci. The capsular poly-saccharide of Friedländer bacilli not only proved to be the carrier of immunological specificity, but turned out to exhibit a close chemical resemblance to the capsular polysaccharide of the type II pneumococcus. This finding immediately suggested a new experiment, which is explicitly stated in the annual report for 1924–25:

> Because the two specific substances, although of widely different biolog-ical origins, resembled each other so closely in some of their chemical properties, the Friedländer polysaccharide was tested with type II anti-pneumococcus serum and a precipitin reaction was found to occur. On the other hand, there was absence of precipitation when this substance was tested with antipneumococcus serum of the other two types.[12]

Even more spectacular was the discovery that mice could be protected against type II pneumococcal infection by treatment with the serum of rabbits which had been vaccinated with encapsulated Friedländer bacilli and that, vice versa, mice could be protected against Friedländer infection by treatment with type II antipneumococcal serum. Clearly, then, immu-nological relationship was not the consequence of biological derivation, but instead was determined by chemical configuration of the capsular substance.

The widespread occurrence of specifically reacting polysaccharides among microorganisms of different biological groups suggested that such polysaccharides might also be found among plants. Indeed, chemical frac-tionation of gum arabic yielded fractions that reacted with type II anti-pneumococcal serum and, to some extent, with type III, but not with type I.[13] Furthermore, immunological relationships were found between the pneumococcus type III polysaccharide and certain plant pectins.[14]

The atmosphere of intellectual excitement created by these discoveries can be recaptured from the style of the reports to the Board of Scientific Directors for the period 1924–25. Some of the passages are worth quoting, because they forcefully express Avery's belief that protective immunity is the expression of the response that the body makes, not so much to a certain microbial species as to a particular molecular configuration, irre-spective of the biological origin of the material possessing the configura-tion. After recalling "the surprising fact that the analogous specific soluble substance of such closely related organisms as the three types of Pneumo-coccus should be so strikingly different," Avery found it "equally remarka-ble that the analogous specific soluble substances of such widely different organisms as this strain of the Encapsulatus group [the Friedländer bacil-lus] and the type II pneumococcus should be so similar." [15] Commenting on the chemical differences between the Friedländer and the type II pneumococcus capsular polysaccharides, he stated further, "it seems reasonable to assume that both contain in a portion of the complex molecule the same or a closely similar steric configuration of atoms. This essential similarity in molecular group would then determine the immunological similarity of the two substances." [16] Thus, the origin of immunological specificity was pushed back from the microbial species to certain molecules that it contains, and from the molecule to the steric configuration of a particular group within the molecule. As will presently be discussed, this concept was soon to be converted into fascinating experimental models by the synthesis of artificial antigens.

Immunity from Sawdust and Egg White

In his speech of acceptance of the Landsteiner Avery Award in 1973, Walther Goebel referred to the 1920s and 1930s as "The Golden Era of Immunology at The Rockefeller Institute." [17] And for good reason. Karl Landsteiner had become a member of The Institute staff in 1922, the very year that Heidelberger joined with Avery in the immunochemical program that was to be so brilliantly developed by Goebel himself. Although both the Avery and Landsteiner departments were intensely involved in immunochemistry, their approaches to the field were at first rather different. Until 1930, Avery and his group were primarily concerned with immunological phenomena as they occur in *natural* systems, especially in infectious processes. In contrast, Landsteiner emphasized immunochemical reactions in *artificial* systems using, instead of bacteria or their products, antigens that he synthesized to elicit antibody production. After 1930, however, these approaches were integrated in Avery's department, largely through Goebel's imaginative immunochemical studies with artificial antigens that had activities similar to those of bacterial products.

While working as a pathologist in Vienna around 1920, Landsteiner produced artificial antigens by combining simple molecules with proteins. Whereas the simple molecules by themselves could not elicit the production of antibodies when injected into animals, they acquired this ability after having been combined with proteins. Landsteiner coined the word "hapten" as a general term to designate substances which cannot act as antigen (i.e., cannot elicit the production of antibodies) by themselves, but acquire this ability when they are part of a larger molecular complex. The capsular polysaccharides of pneumococci fitted well into Landsteiner's hapten concept, since they reacted avidly with sera prepared by immunizing animals with whole pneumococci, yet by themselves were incapable of eliciting the production of antibodies in horses and rabbits.

Taking their lead from Landsteiner's work, Avery and Goebel started, about 1930, a long line of exquisitely designed experiments, in which they synthesized artificial antigens made up of proteins combined with simple sugars, sugar derivatives, or the capsular polysaccharides of pneumococci. By immunizing animals with such artificial antigens, they obtained sera that enabled them to determine the effect of each chemical group on immunological specificity and, thus, to understand the molecular basis of both the immunological differences between pneumococcal types and the immunological similarities between type II pneumococcus and the Friedländer bacillus. They could thus confirm by the methods of synthetic chemistry the assumption made in the 1925–26 report that substances of different biological origin would exhibit immunological similarity if they had the same molecular configuration.

While it would be out of place to review here these very specialized studies, it seems worthwhile to present in Appendix III a simplified account of them prepared by Avery in 1930–31[18] for the lay members of The Rockefeller Institute Board of Trustees. The very fact that he was asked to prepare this account is evidence of the interest aroused by the demonstration that it is possible to prepare at will, by chemical means, complex substances (antigens) that have immunological properties similar to those of pathogenic bacteria.

In later experiments, Avery and Goebel synthesized an artificial antigen containing the sugar derivative cellobiuronic acid, a substance which they knew had a close chemical similarity to the capsular polysaccharides of types III and VIII pneumococci. The immune serum prepared against this synthetic antigen reacted not only with cellobiuronic acid, but also with the various pneumococcal polysaccharides; furthermore, it protected mice against infection with certain types of pneumococci. After describing these spectacular experiments in his book *The Specificity of Serological Reactions,* Karl Landsteiner added that this was "the first time an immune serum produced by means of a synthetic substance acted upon a natural antigen and protected against an infectious disease." [19] The production of immunity against a microbial infection with a synthetic antigen is, unquestionably, the most startling and finest intellectual achievement of medical immunology.

The Avery and Goebel experiments on synthetic antigens found their way into one of the popular weekly magazines under the intriguing title "Immunity from Saw Dust and Egg White." There was more scientific truth than appeared at first sight in this eye-catching headline. In theory, sawdust could be used as raw material for the synthesis of a variety of molecules with the same immunological specificity as the capsular structures of pathogens. These molecules, in turn, could be combined with the protein egg white to produce artificial antigens and, finally, these artificial antigens could be used to elicit immunity against certain infectious diseases. In addition to sawdust and egg white, what is needed to develop a rational approach to immunization based on the use of synthetic antigens is a group of immunologists and chemists sophisticated enough to devise the proper antigen for each particular infectious agent. That is the way of the future.

Biological Specificity

The title of this chapter, "The Chemical Basis of Biological Specificity," is obviously far too sweeping and may even appear misleading, because the text is focused on a limited topic—the role of polysaccharides in immunological specificity. However, the very limitation in the range of subjects discussed, when contrasted with the breadth of their implications, conforms well to Avery's style in scientific research.

During the 1930s, several laboratories were investigating the molecular basis of the specificity of the effects exerted by hormones, vitamins, or enzymes. At The Rockefeller Institute, for example, Max Bergmann and his group of organic chemists were trying to define, in terms of chemical configuration, the lock-and-key image that Emil Fisher had used to describe the specific relationship between an enzyme and the substance it attacks. Avery could have discussed the results of his immunological studies within the larger concepts of specificity that were emerging from the analysis of such biochemical systems, but this was not his bent; he stuck to his lathe. Goebel's discovery in 1925 that two organisms as biologically different as type II pneumococcus and the Friedländer bacillus produced polysaccharides with the same immunological specificity, pointed to the fact that both polysaccharides contained "in a portion of the complex molecule the same or a closely similar steric configuration of atoms." [20] Obviously, Avery could have extrapolated from this observation to other types of biochemical phenomena. Instead, he limited the discussion of its significance to the field of bacterial immunology, using his new insight to undertake the synthesis of artificial antigens that had some of the immunological properties of capsular polysaccharides produced by bacterial pathogens.

This limitation was self-imposed, and did not mean that his experience with polysaccharides in immunological phenomena had made him blind to the role of other kinds of chemical compounds in biological specificity. His openness of mind in this regard can be illustrated by examples taken from his own work, which proved that proteins or nucleic acids are responsible for the specificity of certain biological phenomena.

Early in the 1920s, he and Heidelberger separated from the cells of pneumococci an immunologically active protein that was chemically unrelated to the capsular polysaccharides, and differed from them in other respects. Whereas the capsular polysaccharides are specific for each pneumococcal type, the protein fraction is immunologically the same in all types. However, it is different from protein fractions separated by similar chemical methods from other bacterial species. Thus, Avery himself established that a protein determines the immunological specificity of *Diplococcus pneumoniae*.

The widespread occurrence of hemolytic streptococci in the pneumonia associated with viral influenza during and after the first World War led Avery to become involved in the identification of these organisms. Although his name appears on only two papers concerned with streptococci (in 1919), his views and advice influenced profoundly the classic contributions that Dr. Rebecca Lancefield and her associates made to streptococcal immunology. As mentioned earlier, he had made it a strict policy not to be listed as co-author of a publication unless he had participated in the experimental work with his own hands. Dr. Lancefield, however, would be the first to acknowledge the crucial role of his constant advice in the studies through which she demonstrated that the specificity of group A streptococci is determined by proteins, and not by polysaccharides.

From his early days as bacteriologist to the Hospital, Avery had been interested in antibodies, and had accepted the general view that they are proteins belonging to the group of serum globulins. He published a minor paper on this topic in 1915, when he was trying to concentrate the active fraction of antipneumococcal serum for therapeutic use. Although he did not continue to work actively on the problem, he frequently talked about it in the laboratory and encouraged Goebel, who, in collaboration with John H. Northrop,[21] crystallized the protein antibodies. Avery hoped that the availability of such materials in a pure form would help to identify the steric groups involved in specific reaction with the relevant antigen, as has been done to explain the relationship between enzyme and substrate. These studies did not go far, in part for lack of time, in part because protein chemistry was not then sufficiently advanced for such an analysis, but they provide further evidence that he was not blind to the role of proteins in immunological specificity.

Finally, there is the fact, to be discussed at length in Chapter Eleven, that he devoted an immense amount of effort to the identification of the substance responsible for the transformation of immunological types in pneumococci. He ended his professional life with the demonstration that this genetic phenomenon is not brought about by polysaccharides or proteins, but by deoxyribonucleic acids endowed with specificity.

Thus, his work on specificity ranges over three very different classes of chemical compounds and two classes of biological phenomena. In the laboratory, his experimental material was the pneumococcus but, in his mind, the findings were incorporated in a much broader concept of biological specificity.

CHAPTER NINE

THE COMPLEXITIES OF VIRULENCE

Virulence in Nature and in Experimental Models

Avery, who was so careful in his use of the English language and so meticulous in the expression of his thoughts and feelings, was paradoxically rather casual, at times to the point of carelessness, when it came to the scientific jargon of medical microbiology. Whereas all modern textbooks discuss at length the concept of virulence in an attempt to define the many different shades of meaning the word conveys, he was prone to use it conversationally in its narrow etymological sense, which refers only to the severity of disease. At times, he would speak of virulence as did the physicians of the pre-microbiological era, who could not possibly have known that, when the word is applied to microbial pathogens, it subsumes the multiple mechanisms involved in the complex processes of tissue invasion and of pathological disturbances—mechanisms that differ from one microbial strain to another. This linguistic casualness was not due, of course, to lack of intellectual discipline on his part or to the fact that he underestimated the complexity of virulence. It resulted from the dual manner in which he approached scientific problems.

When Avery became interested in a biological phenomenon, he first observed it for the sheer fun of it, as a naturalist. He took delight in watching almost any manifestation of life, including the pathological ones. At this initial phase of observation, he reported what he perceived in a language that appeared nonscientific because it was not analytical. He simply wanted to convey his direct perception of facts and events, just as he apprehended them. For example, when first describing the pathogens recovered from patients with lobar pneumonia or septic sore throat, he called virulent, without further qualification, the pneumococci, Friedländer bacilli, influenza bacilli, or hemolytic streptococci responsible for the lesions and toxic effects in these diseases. The degree of virulence of a given microbe then meant to him its ability to cause disease, mild or severe, under this or that set of conditions. When looking at infectious processes as a naturalist, he enjoyed conveying his interest in picturesque images without concern for precision. He would playfully speak of the pneumococcus as that cunning little fellow which behaves now as a peaceful citizen, then as a vicious character, depending upon the circumstances.

Once he took the problem of virulence to the laboratory bench, however, this playful posture was replaced by a strictly analytical attitude. Instead of studying virulence in its complex manifestations, as they occur in disease under natural conditions, he tried to reproduce limited aspects of it in one or another experimental model of infection where he could control the variables. He thus explored separately the various facets of virulence from a multiplicity of viewpoints in a number of consecutive steps. A review of these steps, taken in a fairly well-defined chronological order from 1916 to 1942, will bring out his continuity of purpose, even though his approach changed considerably in the course of those 25 years.

As mentioned in Chapter Seven, Avery first considered the problem of virulence from a metabolic point of view. He argued that, since the multiplication of bacteria *in vivo* depends upon the operation of their enzymes, virulence implies that their enzymatic equipment is protected against the defense mechanisms of the host. The exploration of this facet of virulence led him first to the hypothesis of antiblastic immunity in 1916, then to a series of studies on bacterial metabolism, and finally to his 1941 speech before the Society of American Bacteriologists.[1]

The discovery of the specific soluble substances of pneumococci in 1917 and of their polysaccharide nature in 1923, encouraged Avery to focus his attention on the antiphagocytic activity of these substances, and of the bacterial capsule of which they are the chief, if not the only, constituents. This facet of the virulence problem is the one he explored most completely, in both its chemical and morphological aspects. He demonstrated that capsular polysaccharides play an essential role in virulence by protecting pneumococci against engulfment and destruction by phagocytes.

Many strains of pneumococci that are fully encapsulated and produce a capsular polysaccharide known to have antiphagocytic activity, nevertheless are not virulent for certain animal species. In other words, while the production of capsular polysaccharide is a necessary condition of virulence, it is not a sufficient condition. About 1925, Avery postulated that the capsular polysaccharide is an effective factor of virulence only when combined with some other component of the bacterial cell; he assumed also that animal tissues contain an enzyme which, although inactive against the polysaccharide itself, "dissociates" it from its cellular combination. In the light of this dual hypothesis, virulence appeared to him to depend upon the ability of pneumococci to resist the process of antigenic dissociation. He termed this resistance "tissue fastness."

Throughout these studies, Avery was impressed by the fact that pneumococcal strains can undergo reversible hereditary changes in virulence. For example, they can go from the encapsulated form, which may be virulent if other conditions are fulfilled, to the nonencapsulated form, which is never virulent, and vice versa. Similarly, a strain can be made to acquire or lose the tissue fastness essential to virulence for certain animal species. Thus, one of the most intriguing aspects of virulence is the independence and reversible variability of its determinant factors.

The last phase of Avery's exploration of virulence was the genetic analysis of its variability. The gene for each determinant of virulence can be taken from one pneumococcal cell and incorporated into another through the techniques that have come to be known as bacterial transformation (see Chapter Eleven). For example, pneumococci that have lost the ability to produce a capsular polysaccharide can be made to recover virulence by providing them with the genetic equipment needed for the production of that substance; pneumococci that produce one certain type of capsular polysaccharide can be made to produce another type; the tissue fastness that confers virulence for a particular animal species can also be transferred from one pneumococcal strain to another.

Thus, whereas virulence once was regarded as a property pertaining to bacterial cultures considered as a whole, Avery emphasized that it depends on the simultaneous operation of several distinctive attributes of the bacterial cells; he demonstrated that virulence can be altered at will, both qualitatively and quantitatively, by manipulating each of these attributes separately.

Avery's analytical concept of virulence was derived chiefly from his extensive studies of pneumococcal infections, but he applied the same analytical approach to a few other bacterial species, either through his own work or by advice to his colleagues. The most spectacular and best known of his discoveries relating to virulence is that polysaccharides, not proteins, are the substances responsible for the immunological specificity and the antiphagocytic activity of the capsule in pneumococci and Friedländer's bacilli. Of less obvious, but broader, significance is his demonstration that virulence can be analyzed in terms of other chemically defined structures and components of bacterial cells. This concept had, of course, long been established with regard to diseases in which toxin production is the dominant factor of the infectious process, as in diphtheria. Avery's work revealed that the same concept applies to conditions in which virulence depends chiefly upon the ability of the microorganisms to invade tissues.

He showed, furthermore, that there are profound differences in the chemical nature of the factors responsible for invasiveness, even between closely related bacterial groups. Whereas in pneumococci, for example, immunological specificity and antiphagocytic activity are located in the polysaccharide capsule, the analogous role in hemolytic streptococci of group A is played by proteins (the so-called M substances) on the bacterial surface. In hemolytic streptococci of group C, however, invasiveness depends upon the possession of a hyaluronic acid capsule. Although the factors to which Avery attributed tissue fastness are still unidentified, there is reason to believe that they, too, differ in location and chemical characteristics from one bacterial strain to another.

These different facets of the virulence problem were so clearly set apart in Avery's mind that even his speech mannerisms changed, depending upon the aspect of the problem he wanted to emphasize at any given time. If the conversation was focused on the capsular polysaccharide, he would refer to the pneumococcus as this "sugar-coated microbe," whereas he would compare it to Dr. Jekyll and Mr. Hyde if he wanted to discuss reversible changes in virulence.

In a general report to the Corporation of The Rockefeller Institute in 1930, Avery stated that the efforts of his department were centered on an attempt to construct for the pneumococcus a "precise knowledge of the biological properties peculiar to it and the nature of the protective processes which the animal body develops against it."[2] He had begun this program in 1915, and was still deeply involved in it at the time of his retirement. Yet, much of this work has been forgotten. After the advent of the sulfonamides in the mid-1930s and of other antibacterial drugs in the 1940s, the problems of virulence appeared to be of only remote and esoteric interest to most students of infectious disease. Since some of Avery's contributions to the problem have not reached textbooks dealing with mechanisms of infection, it seems justified to present them here in some detail, at the risk of repetition, in part because of their historical interest, and also because they may once more become of practical interest if chemotherapy does not continue to fulfill its promises.

The Bacterial Capsule and Virulence

Ever since 1881, bacteriologists have known that pneumococci recovered from human disease or animal tissues are surrounded by a thick, mucoid envelope. Pasteur called this structure an "aureole," but the more generally accepted name is "capsule." Until approximately 1920, there was no precise knowledge concerning either the chemical composition or the biological properties of the capsule, except that it seemed to be related to virulence. The following quotation from The Rockefeller Institute monograph on "Acute Lobar Pneumonia," which was written largely by Avery, states the consensus on this matter around 1917:

> The exact significance of the capsule of pneumococcus is not known. That it may serve as a protective mechanism of the organism and that it may in some way be related to virulence, is suggested by the fact that capsular development is always much more marked when the organism is grown in animal tissues *in which presumptively there is some opposition to its development.* . . . In addition, it seems to be true that the greater the amount of capsular development the less the amount of passive protection afforded by immune serum (italics mine).[3]

These statements could be read as a preview of Avery's scientific program for the rest of his professional life, but there are indications that, at the time the monograph was published, he was still far from having a clear view of the role that the capsule plays in virulence and immunity.

His first original suggestions concerning a possible mechanism for the protective role of the capsule occurs in the paper on antiblastic immunity that he and Dochez published in 1916. As mentioned earlier, one of the hypotheses presented in the paper is that the capsule protects the enzymes located on the membrane of pneumococci against the effect of the serum antiblastic factors. There is probably a remnant of this hypothesis in the remark quoted above that, in animal tissues, the pneumococcus "encounters some opposition to its development." However, this is the last time in Avery's writings that the capsule is linked to the metabolic activities of the pneumococcus.

In the lobar pneumonia monograph, he referred to the specific soluble substances of pneumococci that he and Dochez had just discovered, but he did not relate them to the capsules. He suggested instead that they might be "responsible for the intoxication which attends pneumococcus infection." While he acknowledged that their "toxicity is in no way comparable to that of diphtheria toxin," he felt, nevertheless, that they possess "a degree of toxicity which, exhibited through the course of an infection, might account for the signs of intoxication in lobar pneumonia."[4] The importance of this hypothetical toxicity appeared to him even greater in view of the fact that the specific soluble substances are released into the infected body as soon as the pneumococci begin to multiply. Thus, his second published hypothesis was that the specific soluble substance contributes to virulence through its toxicity.

It is probable that, after these erroneous hypotheses, his understanding of the role of the capsule in virulence finally emerged from a scientific interest that he had acquired at the beginning of his medical life. While still engaged in the practice of medicine, he received, as mentioned earlier, a small grant to study the relationship between phagocytic index in patients and their susceptibility to infection. This experience made him become co-author in 1910, while at the Hoagland Laboratory, of "Opsonins and Vaccine Therapy" (see Chapter Four). He was, therefore, intellectually prepared to imagine that the specific soluble substances contribute to the virulence of pneumococci by rendering them resistant to phagocytosis, and that the protective antibodies in antipneumococcal serum act as opsonins by combining with the soluble substances.

The phagocytosis-opsonin theory of virulence and immunity is not mentioned in Avery's writings until 1923, either in his publications or the annual reports; nor does the word capsule appear in any of these documents. His silence on these matters is probably explained by the fact that all his papers and reports between 1917 and 1923 dealt with the enzymes of pneumococci and their metabolic activities. In contrast, there was a complete change of scientific content in his writings after 1923. As soon as he and Heidelberger began to report their findings on the chemical composition of the specific soluble substances, they referred to them as capsular polysaccharides, and to the capsule as a structure that protects the pneumococci against phagocytosis. Everything had fallen into place. It is certain, therefore, that Avery had developed an integrated concept of the role of the antiphagocytic role of capsular substances before he resumed publication in this field.

There is other unpublished evidence that he had long been interested in the mechanisms that render virulent pneumococci resistant to phagocytosis. In conversation, he referred now and then to some of his early experiments, in which he had tested various chemical compounds for their ability to neutralize the antiphagocytic power of the capsule. For example, he carried out phagocytic tests *in vitro* and protection tests in mice with a variety of mineral salts and organic substances that he found, by microscopic examination, to be capable of reacting with the capsule. He had hoped that the neutralization of antiphagocytic power by chemical substances would open the way for a rational chemotherapy. All these experiments failed, and he did not even mention them in the annual reports, but they testify to the constancy, intensity, and diversity of his interest in the role of the capsule in pneumococcal infections.

The discovery that the specific soluble substances in the capsules are made up of polysaccharides probably made him feel that conditions were now right for a more searching analysis of virulence and immunity. He began his presentation of the problem to the Board of Scientific Directors with an understatement of what he knew and believed: "The synthesis of this polysaccharide is a cellular function highly developed in those strains of pneumococci which are most capable of multiplying in animal tissues. This substance *apparently* bears a significant relationship not only to type specificity but to virulence and capsular development" (italics mine).[5] The word "apparently" means here that, although he was convinced, he did not regard the evidence as final proof until he could illustrate it with one of his demonstrative "protocol experiments" (see Chapter Five). In fact, evidence for the theory of a relationship between capsular polysaccharide and virulence rapidly became so overwhelming that it was accepted as textbook knowledge within a few years.

The theory was clear, but it had disturbing practical limitations. From the beginning of the pneumonia work at the Hospital, highly effective therapeutic sera against type I pneumococci had been consistently obtained by the immunization of horses, but the results had been far less satisfying with other types, and entirely negative with type III. This failure certainly accounted for the interest Avery took in me during the meeting recounted in Chapter Five. He hoped that if one could find an enzyme capable of decomposing the type III capsular polysaccharide, and if this enzyme could be shown to destroy the capsule itself and thereby render the pneumococci susceptible to phagocytosis, his theory would be vindicated. In fact, the results of the enzymatic approach went beyond expectation. The bacterial enzyme that we found to be capable of decomposing the specific polysaccharide of type III pneumococci did not interfere with the growth of pneumococci *in vitro,* but, when it was injected into infected animals, it destroyed the bacterial capsules so rapidly and completely that the phagocytes could immediately engulf the bacteria and kill them. Mice, rabbits, and monkeys suffering from advanced infection with type III pneumococci promptly recovered after treatment with the enzyme. This provided further proof, if any was needed, of Avery's doctrine concerning the role of capsular polysaccharide in virulence.

The Bacterial Body and Virulence

In their purified soluble forms, all the capsular polysaccharides react avidly with the corresponding antibodies of antipneumococcal sera that have been obtained either from pneumonia patients or from animals vaccinated with the killed cells of virulent pneumococci of the proper type.

However, while such cells can act as antigens, none of the capsular polysaccharides is capable by itself of eliciting the formation of antibodies when injected into horses or rabbits; it behaves as a hapten. From these facts, Avery concluded that the capsular polysaccharides exist in the pneumococcal cells as a part of complex structures that confer upon them the antigenicity they do not possess after they have been separated in pure form. The analysis of the factors involved in the antigenicity of the capsular polysaccharides was for many years one of his most constant preoccupations.

The point of departure of the analysis was the observation that, when encapsulated pneumococci are allowed to disintegrate by autolysis before being used as vaccines, they do not elicit the formation of antibodies against their capsular polysaccharides, even though these substances persist in the autolysate. Avery assumed that the polysaccharide was separated during autolysis by some pneumococcal enzyme from the hypothetical cellular structure which endows it with antigenicity—what he called the complete capsular antigen. He referred to this separation or splitting as "antigenic dissociation." The tables of contents of the annual reports for 1930–1931 and 1936–1937 give an idea of the enormous amount of experimental work that was devoted for several years to this hypothesis (Appendix IV).

The program on antigenic dissociation involved extensive studies of the autolytic system of pneumococci and of the effects exerted by various enzymes on the antigenic activities of the bacterial cell. The results were disappointing from the theoretical point of view, because they did not elucidate the chemical composition of the complete capsular antigen or the mechanisms of antigenic dissociation; nevertheless, they had great practical utility. Despite this failure, it was still possible to develop empirical means that prevented, or at least minimized, autolytic processes. This led to the production of highly effective vaccines for the preparation of therapeutic sera. By 1935, as mentioned in Chapter Eight, successful results were obtained in the Hospital with serum treatment of many types of pneumococcal lobar pneumonia.

Even with the best vaccines, however, it was more difficult to obtain sera containing a high level of specific antibodies against type II and, especially, type III pneumococci than against type I. To explain this anomaly, Avery postulated that there exist in the animal body certain enzymes capable of splitting the complete capsular antigen, and that this splitting occurs more rapidly in type II and type III than in type I. Here, again, he referred to the hypothetical splitting as antigenic dissociation, but in this case brought about by the enzymes of animal tissues, instead of by the pneumococcal enzymes.

Carrying his speculations still further, he suggested that "the factors which make for dissociation of the antigen after injection into the animal body . . . appear to be related to what is commonly called natural immunity, for animals which are most resistant to pneumococcus infection are just those animals which have been found to possess the greatest capacity to split the antigen and consequently to yield the least potent serum."[6] According to this mental construct, certain animal species are endowed with natural resistance to a given type of encapsulated pneumococcus because they possess an enzymatic machinery that rapidly dissociates the complete capsular antigen of this particular type. In the light of this hypothesis, the difficulty experienced in obtaining sera with a high level of antibodies against type II and type III pneumococci came from the fact that the complete capsular antigens of these types were rapidly dissociated, either by the pneumococcal autolytic enzymes or by the enzymes of the animal tissues.

Avery thought at first that the differences in antigenic stability of the various pneumococcal types could be traced to chemical differences in the capsular polysaccharides themselves:

> The fact that this splitting of the specifically immunizing complex of pneumococci occurs so readily, particularly in the case of organisms of Type II and III and the fact that under these circumstances the stimulus to specific antibody production is lost so quickly in these two instances affords a possible explanation of the lack of success in obtaining an antiserum of high potency against these types. It seems not unlikely that the relative differences in the rate and degree of splitting of the specific antigens in the three types of pneumococcus are in each instance referable to *known differences in the chemical structures of the specific sugar components.* Antigenic stability, like specificity itself, then rests upon the chemical constitution of these unique and specific substances (italics mine).[7]

New findings, however, revealed that this hypothesis was erroneous. Some strains of type III pneumococcus were found to be virulent for rabbits and others not, even though they produced equally large amounts of the same capsular polysaccharide. Even if it were true that the virulence characteristics of the two strains could be explained by the rates of antigenic dissociation, the difference between them was obviously due to some factor other than the polysaccharide itself, since this was the same in both cases.

Instead of being discouraged by the finding, Avery formulated a further hypothesis. He assumed that certain strains of type III pneumococci had become virulent for rabbits through a change that had rendered their complete capsular antigen more resistant to the enzymes of the rabbit. He imagined that the "rabbit virulence factor" was the cellular expression of some kind of tissue fastness—to use his own words—that had its site not in the capsule, but in the body of the pneumococcus; it was a "somatic" factor. He assumed also that the mechanism responsible for the degree of virulence in some way determined the ability of the culture to elicit in rabbits the formation of antibodies specific against type III pneumococci.

It is probably impossible for anyone who was not a member of Avery's department during the late 1920s and the 1930s to follow the succession of hypotheses he formulated to explain in immunological and enzymatic terms the complexities of the virulence problem—hypotheses that were more remarkable for their imaginative exuberance than for their value as guides to experimentation. An idea of the multiple experimental approaches that were developed in the department during that period can be had from the table of contents of the annual report for 1926–27 (Appendix V).

Despite extensive experimentation by most of us, the findings were never sufficient to evaluate the validity of the hypotheses formulated to explain antigenic dissociation or to isolate the somatic virulence factor. For this reason, the only aspects of the annual reports that found their way into scientific journals are those that yielded clear-cut experimental facts with a straightforward interpretation, independent of any hypothesis concerning the mechanism of virulence. For example, although the phrase "tissue fastness" occurs repeatedly in the annual reports, and although I have used it in these pages, I do not believe that it occurs in any of the papers published from the department.

Avery's failure to identify the nature of the rabbit virulence factor did not cause him to lose interest in the problem. In fact, he came back to it in the 1940s when the techniques of bacterial transformation (Chapter Ten) made it possible to put to the test his hypothesis that, in pneumococci of type III, virulence for rabbits depends upon a somatic factor completely independent of the capsular polysaccharide. Although the principle of the "transformation" technique will not be presented until the next chapter, it seems worthwhile to describe this particular experiment here, because it reveals that one of the determinants of rabbit virulence persists, in an unexpressed form, in certain nonvirulent pneumococci.

Two strains of type III pneumococci, both fully encapsulated, but one virulent for rabbits and the other nonvirulent, were caused to undergo a genetic change that resulted in the loss of their ability to produce a capsule. After being deprived of the capsule, even the strain that had been rabbit-virulent was incapable of causing disease. The two nonvirulent cultures were then converted into type II pneumococci by treatment with the transforming material prepared from these organisms. When the cultures of type II pneumococci thus artificially produced by the transformation technique were tested in rabbits, it was found that they differed in virulence — their ability to establish disease in rabbits corresponded to that of the particular type III strain from which they had been derived; one was virulent for rabbits and the other was not. The importance of this finding is the demonstration that one of the determinants of rabbit virulence is associated with a cellular factor which persists in the noncapsulated form, independent of the production of capsular polysaccharide.

The experimental feat just described was performed by MacLeod and McCarty and published by them in 1942 under the title "The relation of a somatic factor to virulence of pneumococci."[8] Avery did not want to have his name entered as co-author of the article because he had not actually participated in the experiments, but the genetic transfer of the rabbit virulence factor was obviously the experimental demonstration of the mechanism he had postulated 10 years before under the name of tissue fastness. It is not unlikely that, in a similar way, his mental constructs about antigenic dissociation will acquire a concrete meaning if techniques ever become available to determine how the capsular polysaccharide is bound in the intact cellular structure of pneumococci.

CHAPTER TEN

BACTERIAL VARIABILITY

Polymorphism vs. Monomorphism

During the two centuries that elapsed after Leuwenhoek first saw bacteria, probably about 1675, these organisms were studied almost exclusively by microscopic examinations; naturalists were primarily interested in their occurrence, shapes, and motility in different natural fluids and products, or in the bodies of human beings and animals. The paper on lactic acid fermentation published by Louis Pasteur in 1857 is the first well-documented report of a study in which a bacterial culture was manipulated under controlled conditions to measure its chemical activity, rather than to observe its morphological appearance.[1] Pasteur was intensely interested in what bacteria do and in the specificity of their chemical and pathological effects, but he paid little attention to their cellular organization and other purely biological characteristics.

Until the middle of the nineteenth century, in fact, most microscopists believed that bacteria were extremely primitive organisms, so simple as to be little more than poorly organized chunks of protoplasm. "They form the boundary line of life; beyond them, life does not exist," wrote the botanist Ferdinand Cohn in a short, classic essay published in 1866 under the title "Ueber Bacterien, die kleinsten lebenden Wesen."[2] Such assumed simplicity of structure led many biologists of the time to believe that the various bacterial forms seen under the microscope were but the different manifestations of only one or a very few elementary protoplasmic structures that could change in appearance and other characteristics, depending upon environmental conditions. This now-discredited theory, which has been called the doctrine of bacterial polymorphism, was asserted in one form or another by many of the most famous biologists and physicians until a century ago—for example, by Thomas Huxley in 1870,[3] by Edwin Klebs in 1873,[4] by Ray Lankester in 1873,[5] by Theodor Billroth in 1874,[6] and even by the illustrious surgeon Joseph Lister in 1873 and 1876.[7] Louis Pasteur and Ferdinand Cohn were among the very few scientists who explicitly rejected the theory and who believed from the beginning in the distinctiveness and biological stability of the different bacterial types.

An elaborate statement of the doctrine of bacterial polymorphism was published in 1877 by the botanist Carl von Nägeli in his book *Die Niederen Pilze*,[8] in which he introduced the word *Anpassung* (adaptation, acclimatization) to express his view that bacteria were primitive fungi capable of changing from one morphological or physiological type to another as they adapted to new external conditions. Nägeli's thesis had a peculiar fate. Even before it was published, bacterial polymorphism was being abandoned and replaced by an opposite doctrine of strict bacterial monomorphism. Yet the concept of *Anpassung* survived and was used extensively a few years later to account for the discovery that bacterial species do, in fact, continuously give rise to many variant forms in response to environmental changes.

From the time of his first biological studies in 1857, however, Pasteur believed that, for each kind of fermentation and each kind of contagious disease, there exists a particular type of microorganism that retains its fundamental characteristics under all conditions, but he could not provide biological evidence to prove his point. Having been trained in physics and chemistry, and having only limited knowledge of conventional biology, he could not give morphological descriptions of the microorganisms he studied; instead, he put his emphasis on their functional attributes, such as their ability to perform certain chemical reactions or to cause certain pathological disorders. Evidence for the distinctiveness of bacterial types required the use of biological criteria such as those introduced in the 1870s by Ferdinand Cohn and by Robert Koch.

Many different scientific forces, acting over several decades, played roles in discrediting the doctrine of bacterial polymorphism. They can be symbolized by three very different types of studies that were published during 1876, each contributing in its own way to the demonstration that bacteria are well-defined biological entities, stable in their fundamental characteristics.

In 1876, Ferdinand Cohn published the fourth of his *Untersuchungen über Bacterien*,[9] in which he gave precise descriptions of the morphological appearance of various bacterial cells as seen under the microscope, and suggested that they could be classified in four morphological groups, each consisting of several genera. He thus introduced into bacteriology taxonomic criteria similar to those used in other biological fields. In 1876, Koch published *Die Aetiologie der Milzbrandkrankheit*,[10] in which he described the life history of the anthrax bacillus and its role as an agent of disease. Also in 1876, Louis Pasteur published his *Etudes sur la bière . . . avec une théorie nouvelle de la fermentation*.[11] In this work, he presented in their final forms the views on the biological specificity of fermentative processes and on other chemical aspects of microbial life that he had been expounding for some 20 years.

The distinctiveness and stability of bacterial species—the so-called doctrine of bacterial monomorphism—had thus been fully demonstrated by 1876, or so it was believed. There is no doubt that the very rigidity of this doctrine helped to create bacteriological science by introducing discipline into the intellectual approach to problems and into the design of techniques. The subsequent triumphs of the germ theory of fermentation and disease would not have been possible without this discipline. As we shall now see, however, the doctrine of monomorphism did not last long in its rigid, original formulation.

Phenotypic Adaptations and Hereditary Changes

Within a decade after the doctrine of bacterial polymorphism had been discredited, many bacteriologists became aware that each bacterial species can undergo profound changes in many of its characteristics. Practical and theoretical developments soon emerged from the recognition that there exist many variant forms within a given species.

Ironically, Pasteur, who had been the first to affirm that bacterial species are distinct and stable, was also the first to recognize bacterial variability during his work on fermentation and infection. Furthermore, he came to regard bacterial variability as a manifestation of the adaptive phenomena that Carl von Nägeli had designated by the word *Anpassung.* Nägeli's interpretation was faulty, because what he thought to be transformation of one species into another was, in reality, a succession of different bacterial species in mixed cultures. It turned out that the word *Anpassung* was being used for two very different mechanisms of change within a given species.

In some cases, *Anpassung* denotes the rapid and reversible changes in morphological appearance and physiological behavior that a particular organism can undergo by phenotypic adaptation in direct response to a change in its environment. In the *Etudes sur la bière,* for example, Pasteur had shown that, whereas yeast cells are globular during fermentation, they become somewhat elongated in the presence of oxygen; conversely, the fungus *Mucor,* which usually grows as a mycelium, becomes globular and yeastlike under anaerobic conditions. Pasteur also showed that the amounts of CO_2, alcohol, organic acids, and protoplasmic material produced by yeast from a given amount of sugar differ greatly according to the oxygen tension in the culture medium. Although such morphological and metabolic changes can be profound, they are not lasting; they correspond to a phenotypic response of the individual organism, and are not transmissible to its descendants.

In other cases, *Anpassung* denotes adaptive changes which, in contrast to phenotypic adaptations, are hereditary. Examples of such hereditary adaptive changes among bacteria were first observed by students of experimental infections.

As early as 1872, three years before Koch and Pasteur had published anything on the anthrax bacillus and before the pathogenic role of this organism had been established, C. J. Davaine had discovered that the "virulence" of the blood of rabbits infected with anthrax could be spectacularly increased by the technique he introduced under the name "animal passage," namely, the inoculation of animals in series with smaller and smaller amounts of infected blood. Davaine observed that the causative agent progressively increases in activity as it passes through living animals ("acquiert donc une plus grande activité en passant par l'économie d'un animal vivant").[12] Shortly after he began to work with bacterial diseases, Pasteur himself postulated that bacteria become more virulent by animal passage because they undergo a process of biological adaptation to the animal through which they are "passed." Referring to fowl cholera, he postulated that its bacterial agent "having multiplied for numerous generations [in the bodies of chickens] becomes more and more capable of overcoming their natural resistance in the same way as different races of animals and human beings progressively become acclimatized to a new environment." The adaptive changes thus produced by animal passage are fundamentally different from phenotypic adaptation because they are transmissible from one bacterial culture to the next.

In 1880, Pasteur observed accidentally that the causative agent of fowl cholera commonly becomes nonvirulent when cultivated *in vitro*; he referred to this phenomenon as "attenuation" of virulence.[13] In 1881, he developed techniques by which he could at will cause cultures of the anthrax bacillus to lose the ability to produce spores and simultaneously to lose virulence for cattle.[14] He used attenuated forms of these two bacterial species to develop vaccines that conferred immunity to fowl cholera and to anthrax respectively, thereby opening a general approach to vaccination by means of living, attenuated cultures.

Because of their relevance to disease, the changes resulting in exaltation or attenuation of virulence were the two kinds of hereditary variation in bacteria that were at first most widely recognized and most extensively studied. However, other kinds of hereditary changes also were described by the early bacteriologists. In 1888, for example, G. Firtsch recognized

1 · *The Reverend and Mrs. Joseph Francis Avery and sons Ernest (standing), Oswald (seated), and infant Roy, about 1886.*

2 · *Mariners' Temple at 1 Henry Street, where J. F. Avery was pastor from 1887 until his death in 1892, is still standing.*

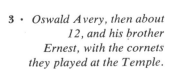

3 · *Oswald Avery, then about 12, and his brother Ernest, with the cornets they played at the Temple.*

4 · Avery, seated third from left, with the Colgate University band.

5 · *Goats browse on farmland that became the site of the Institute.*

6 · *Below, the Institute in 1913, the year Avery joined the staff of the Hospital, at right.*

7 · *First board of directors of the Institute. Left to right: T. Mitchell Prudden, Christian A. Herter, L. Emmett Holt, Simon Flexner, William H. Welch, Hermann M. Biggs, Theobald Smith.*

THE LABORATORIES OF
THE ROCKEFELLER INSTITUTE FOR MEDICAL RESEARCH,
66TH STREET AND AVENUE A, NEW YORK.

Chocorua, N.H.,
August 15, 1913.

Dear Dr. Prudden:

      The Hospital position of Bacteriologist has not been filled this year, and Dr. Cole recommends Dr. Oswald T. Avery, a brief history of whom is enclosed. I also know Dr. Avery, and I strongly recommend him for the appointment. It takes effect on September 1, 1913, and expires July 1, 1914, the salary being at the rate of $2000 per year.

      Yours very truly,

Simon Flexner

I approve the appointment of Dr Avery
T. M. Prudden

Aug 19th 1913

8 · *Appointment of Avery to the Hospital was confirmed in note, above.*

9 · *Culture of pneumococcus isolated and reported by Avery in 1916, left, gave rise to strains used in the discovery that DNA carried genetic information.*

10 · *A portion of the letter Simon Flexner received in 1932 from Avery, who was suffering from Graves' disease.*

Deer Isle, Maine

Dear Dr Flexner: —

Not another day shall pass without my sending a little note to express my appreciation of your kindly thought [in] to assure you of my improvement.

The rest and change during these delightful days on the Maine coast have done me a great deal of good — I have gained in weight and feel better than I have for

11 · *Avery after being commissioned captain in the U.S. Army during World War I.*

12 · *The Hospital mellowed over the years as ivy began to cover its stark, utilitarian lines.*

13 · *Members of Avery's department about 1932. Left to right: seated, Thomas Francis, Jr., Avery, Walther F. Goebel; standing, Edward E. Terrell, Kenneth Goodner, René J. Dubos, Frank H. Babers.*

14 · MICHAEL HEIDELBERGER

15 · WILLIAM H. WELCH

16 · RUFUS COLE

17 · A. R. DOCHEZ

to look forward to happier days even if they are not
perfect — we can take it — & still be happy —

Forgiving this rambling epistle — with it goes
my love & thoughts & the hope of better things ahead —

"With heaps & heaps of love"

Affectionately & Faithfully

O.M.

If the Board in the Surgeon General's office
meets at Camp Bragg and then they may later on
I shall take the opportunity of running over to Nashville
for I want to take over future plans & possibilities
with you & Catherine — Do write if just a line —
I want to know your reaction — & don't hesitate
to talk to Ernest — he knows it all & we talked
it over very frankly —

Good night — it's long after mid-night — & I
have a busy day ahead — God bless us, one & all
sleepy — Will & Harry —

18 · *In this "rambling epistle" to his brother Roy, Avery described the pneumoccal transformation studies. The complete text is in Appendix I.*

STUDIES ON THE CHEMICAL NATURE OF THE SUBSTANCE
INDUCING TRANSFORMATION OF PNEUMOCOCCAL TYPES

INDUCTION OF TRANSFORMATION BY A DESOXYRIBONUCLEIC ACID FRACTION
ISOLATED FROM PNEUMOCOCCUS TYPE III

By OSWALD T. AVERY, M.D., COLIN M. MacLEOD, M.D., AND
MACLYN McCARTY,* M.D.

(From the Hospital of The Rockefeller Institute for Medical Research)

PLATE 1

(Received for publication, November 1, 1943)

Biologists have long attempted by chemical means to induce in higher organisms predictable and specific changes which thereafter could be transmitted in series as hereditary characters. Among microörganisms the most striking example of inheritable and specific alterations in cell structure and function that can be experimentally induced and are reproducible under well defined and adequately controlled conditions is the transformation of specific types of Pneumococcus. This phenomenon was first described by Griffith (1) who succeeded in transforming an attenuated and non-encapsulated (R) variant derived from one specific type into fully encapsulated and virulent (S) cells of a heterologous specific type. A typical instance will suffice to illustrate the techniques originally used and serve to indicate the wide variety of transformations that are possible within the limits of this bacterial species.

Griffith found that mice injected subcutaneously with a small amount of a living R culture derived from Pneumococcus Type II together with a large inoculum of heat-killed Type III (S) cells frequently succumbed to infection, and that the heart's blood of these animals yielded Type III pneumococci in pure culture. The fact that the R strain was avirulent and incapable by itself of causing fatal bacteremia and the additional fact that the heated suspension of Type III cells contained no viable organisms brought convincing evidence that the R forms growing under these conditions had newly acquired the capsular structure and biological specificity of Type III pneumococci.

The original observations of Griffith were later confirmed by Neufeld and Levinthal (2), and by Baurhenn (3) abroad, and by Dawson (4) in this laboratory. Subsequently Dawson and Sia (5) succeeded in inducing transformation *in vitro*. This they accomplished by growing R cells in a fluid medium containing anti-R serum and heat-killed encapsulated S cells. They showed that in the test tube as in the animal body transformation can be selectively induced, depending on the type specificity of the S cells used in the reaction system. Later, Alloway (6) was able to cause

---

* Work done in part as Fellow in the Medical Sciences of the National Research Council.

Discussion —

The present study deals with the isolation of desoxyribose nucleic acid from Type III pneumococci with demonstration of its activity in inducing specific structural and functional changes in unencapsulated R forms derived from a culture of Pneumococcus Type II.

So far as the workers are aware this is the first time that desoxyribo-nucleic acid (thymus type) has been recovered from pneumococci and the only instance in which specific transformation has been experimentally induced in vitro by a chemically defined substance.

Although the observations are limited to a single example, they acquire broader significance from the work of earlier investigators who demonstrated the interconvertibility of various pneumococcal types and showed that the specificity of the changes induced is in each instance determined by the particular type of encapsulated cells used to evoke the reaction. From the viewpoint of the phenomenon in general, therefore, it is of special interest that in the example cited highly purified and protein-free material consisting largely of desoxyribonucleic acid stimulating unencapsulated R variants derived from Type II pneumococci to produce a capsular polysaccharide identical in type specificity to that of the cells from which the inducing substance was isolated.

20 · *Avery's handwritten text of the discussion section of the paper on the facing page.*

21 · *Colin M. MacLeod, left, and Maclyn McCarty at the dedication of the Avery Memorial Gateway at The Rockefeller University on September 29, 1965.*

colonial variation in *Vibrio*[15]; in 1901, Martinus Beijerinck published photographs illustrating a wide range of variations in cellular and colonial morphology.[16] Max Neisser in 1906[17] and R. Massini in 1907[18] described the spontaneous appearance of lactose fermenting organisms in cultures of the nonlactose fermenter *Escherichia coli mutabile*. Similar examples of hereditary changes in biochemical properties were soon reported in the scientific literature and made it obvious that phenomena akin to mutation are common in bacteria.

In 1921, there came to light a morphological expression of bacterial variability that was destined to have a deep influence on medical bacteriology and, later, on genetics. While comparing the virulent strains of Shiga dysentery bacilli with the nonvirulent forms that appear spontaneously in laboratory cultures of this species, the British bacteriologist J. Λ. Arkwright noticed that the colonies of the virulent forms are regular in shape, dome-shaped, and with a smooth surface, whereas the colonies of nonvirulent forms are irregular, granular, and flat.[19] He introduced the terms Smooth and Rough (respectively abbreviated as S and R) to describe the colonial appearance of the culture, then eventually to denote the culture itself. He postulated that persistent variations in virulence and colonial morphology correspond to genetic mutations.

Arkwright also noticed that R forms occur frequently in cultures grown under artificial conditions, but not in infected tissues. From this he concluded that a Darwinian process of natural selection is at work in determining whether S or R forms gain the upper hand. In his words, "The human body infected with dysentery may be considered a selective environment which keeps such pathogenic bacteria in the forms in which they are usually encountered."[20] The importance of Arkwright's discoveries and interpretations was immediately recognized, and several authors in different countries, working with other bacterial pathogens, confirmed that loss of virulence was commonly associated with a change in colonial morphology. The expressions S and R thus came to be associated with both the colonial morphology of the culture and its virulence; the reversible change S⇄R came to be regarded as a process involving mechanisms of mutation and Darwinian selection.

The Many Facets of Bacterial Variability

Mendelian genetics had at first little influence on the scientific climate in which the understanding of bacterial variation progressively emerged during the first half of the twentieth century. By 1900, the original doctrine of bacterial polymorphism had been completely discredited. Bacteriologists were convinced that bacteria breed true like other living things; they tried to classify and name the best-defined organisms according to the Linnean system, using names such as *Lactobacillus bulgaricus* and *Lactobacillus acidophilis; Streptococcus hemolyticus* and *Streptococcus fecalis; Mycobacterium tuberculosis* and *Mycobacterium bovis.* Complications appeared now and then, but they did not affect the belief that, on the whole, microorganisms are stable biological entities possessing distinctive functional attributes.

However, even the most orthodox and conservative among bacteriologists could not help noticing that the doctrine of bacterial monomorphism failed to account for a multiplicity of variant forms that constantly appear under natural conditions, as well as in the laboratory. Even pure cultures — clones — issued from single cells commonly undergo modifications in some of their morphological, biochemical, and pathological attributes. Such modifications appear spontaneously or can be brought about deliberately by experimental procedures; some persist in subcultures and thus seem to be hereditary, whereas others are extremely transient. As the phenomenon of bacterial variability was observed under a great variety of ill-defined conditions without any understanding of its mechanisms, it gave rise to a confusing terminology that reflected a confusion of thought in the bacteriological community.

Individual bacteriologists, however, had fairly well-formulated views of bacterial variability. Some considered it as the outcome of "training" or "adapting" cultures to new substances or to new animal hosts and, more generally, to new environmental conditions. Others used names such as saltation, Dauermodifikation, phase variation, bacterial dissociation, etc. Still others borrowed the terminology of classical genetics, and referred to mutations, genotypes, and phenotypes.

In most cases, the diversity and complexity of the changes observed, the rapidity with which they occurred, and the ease of their reversibility made it difficult to believe that the classical concepts of genetics sufficed to explain the variability of bacteria. This skepticism was strengthened by the fact that nothing was yet known concerning the existence of nuclei, chromosomes, or genes in bacteria. When I wrote *The Bacterial Cell*[21] between 1943 and 1945, I could find in the literature only a few sketchy experiments to support the view that, despite obvious differences between bacteria and other living things, some phenomena of bacterial variability nevertheless probably fall within the fold of classical genetics; I had to print an addendum to the book, at the page-proof stage, to describe the evidence published in late 1944 that the cells of several bacterial species do indeed possess the equivalent of a discrete nuclear apparatus or, more precisely, a nucleoid body. As was shown later, the cell's DNA is concentrated in this body.

It is now clear that, as used in the past, the expression "bacterial variability" included several kinds of unrelated phenomena; the multitude of names that were used to describe the different kinds of variations can probably be classified under two general headings, phenotypic excursions and genetic mutations. Each of the bacterial species has a wide repertoire of phenotypic expressions affecting its morphological, physiological, and pathological characteristics. The conditions under which the species is grown, and the stage of its development at which it is observed, determine which aspects of its repertoire are expressed or repressed. In the 1930s, for example, certain bacterial enzymes were termed "adaptive" because they were found to be produced in significant amounts only as a specific response to the presence of the homologous substrate in the culture medium.[22-25] Such adaptive enzymes are now called "inducible," and the mechanism of their induction has been traced to the fact that the proper substrate acts by neutralizing the product (repressor) of a regulatory gene that otherwise prevents enzyme synthesis.[26]

Just as interesting as the effect of the composition of the medium on phenotypic expression, but less well known, is that the individual cells of bacterial cultures undergo profound changes in the course of their growth cycles. This topic is discussed at some length in *The Bacterial Cell* under the name "cytomorphosis"[27]; it proved of great importance with regard to the conditions under which bacterial cells acquire the "competence" to incorporate the DNA of another cell and thus undergo transformation.[28]

Except for a few special cases, such as enzyme induction, the development of competence, and the effect of certain substances and conditions on cellular characteristics, the field of phenotypic variations in bacteria has been neglected. In contrast, the transmission of hereditary characteristics has been studied extensively during the past three decades, and has been shown to take place through genetic mechanisms analogous to those operating in other organisms. Unexpectedly, this aspect of bacterial knowledge emerged from the study of the pneumococcus, an organism so delicate that it would have seemed unsuited *a priori* for the complex laboratory operations involved in this kind of study. Also unexpectedly, the understanding of the chemical processes involved in heredity was developed initially not by theoretical biologists or even by general bacteriologists, but by students of infectious processes. Bacterial genetics, which led to modern chemical genetics, emerged in large part from the analysis of

virulence and of colonial variation in pneumococci, first by an epidemiolo-
gist in England and, shortly after, by young physicians working on lobar
pneumonia under Avery's leadership in the Hospital of The Rockefeller
Institute for Medical Research.

Transformation of Types in Pneumococci

Fred Griffith, a medical officer who worked in the pathology laboratory
of the British Ministry of Health, was the first to describe the S and R
forms of pneumococci. Like Avery, whose contemporary he was, Griffith
was a quiet and retiring bachelor, a recluse, known to few. In the words of
one of his students, he was the most English of Englishmen. He was small
and of slight build, with a long, thin face. He spoke with great precision,
but in a whisper. At an international meeting of microbiologists in 1936,
he presented an important paper on the classification of streptococci by
talking down to his manuscript in such a low voice that nobody heard a
word he said. According to one of his American friends, however, he
became a lively companion and a better conversationalist when vacationing
on the Downs near Brighton. He had there a house that looked so modern
for the time (1936) that it shocked his conservative friends. He would walk
briskly on the Downs, and this ultracautious scientist drove his car along
narrow streets at a speed that frightened his American visitor.[29]

Griffith loved bacteriology and epidemiology, but, above all, "He was a
civil servant, and proud of it. He had that kind of a mind and the integrity
that often goes with it. He did not allow his fancy to roam . . . and being
employed by the Ministry of Health to do a specific job, he believed in
fulfilling his contract however frustrating that might be,"[30] and however
limited the working conditions put at his disposal. His laboratory facilities
were limited, indeed, but he "could do more with a kerosene tin and a
primus stove than most men could do with a palace."[31] Especially, he made
up for limitation of equipment by the ingenuity and meticulous care with
which he conducted his experiments. According to Stuart Elliott, who
worked with him for many years, he was "fanatic about techniques," and
wanted his associates to carry out tests exactly according to his teachings.
His meticulousness "sometimes aroused the exasperation, if not the fury,
of his associates and assistants."

Griffith was so retiring that he could rarely be persuaded to attend
scientific meetings, let alone to present a paper, but he was always willing
to help, at whatever cost of time and trouble. "It did not matter if the
visitor were the veriest tyro humbly seeking an introduction to a particular

technique, or a senior of repute wishing to discuss some intricacy of public health bacteriology, all came away impressed" with the compendium of knowledge and the wealth of practical experience that he used in his own work and generously made available to others.[32]

In his scientific life, Griffith followed a single star. He believed that progress in the epidemiology of infectious diseases could come, and only come, with better knowledge of the microorganisms involved, and with better ways of differentiating one strain from another. To this task, he devoted the thirty years of his professional life, quietly accumulating observation after observation, with a clear end in view. He wanted to develop practical and dependable techniques for the identification and classification of pathogenic species. In addition to his work with pneumococci, he introduced a practicable system for typing hemolytic streptococci, and thus made possible the epidemiological study of infections caused by these microorganisms. He was killed in 1941 during an air raid over London, not in a shelter or even indoors, but while "fire-watching." To the end, he had done the job that he felt his duty to perform as a public servant.

Griffith's discovery of the S and R forms of pneumococci was made in 1923.[33] He made the further fundamental observation that, when large numbers of avirulent R cells are injected into mice, it is often possible to recover from the heart blood of these animals S cells which are fully virulent, and which possess a capsular polysaccharide with the same immunological type as the S cells from which the R cells were orginally derived. He concluded from these and other observations that the potentiality for virulence persists in the nonvirulent cultures, and that the animal body acts as a selective medium in which only the S forms can multiply. Step by step, Griffith developed techniques that enabled him to transform at will one colonial form of pneumococci into another, both *in vivo* and *in vitro*. In particular, he found that the R form can be transformed into the S form *in vitro* by adding anti-R immune serum to the culture medium. These experiments led him to postulate that the S & R variation is a reversible mutational change.

While he regarded the reversibility from S to R, and vice versa, as a mechanism by which the organisms adapt themselves to new environmental conditions either *in vitro* or *in vivo*, Griffith took it for granted that the changes remained within the limits of the species. He probably had not envisaged that one pneumococcus type could be transformed into another, as this was then regarded as the equivalent of transforming one species into another—a phenomenon never observed. Yet, such a transformation is precisely what he himself observed in 1928,[34] thus exploding a bombshell in the field of pneumococcal immunology.

The experiment by which Griffith demonstrated that pneumococci can be made to change types deserves to be described in some detail because of its historical importance and because the discovery of the new phenomenon is a striking example of serendipity. It also provides an illustration of the fact that, while progress in science often depends on interesting accidents, these can be recognized and exploited only by investigators endowed with theoretical knowledge and experimental skills. In Pasteur's words, chance favors only the prepared mind.

One of the empirical techniques used by bacteriologists to establish experimental infection in laboratory animals is to inject the infective inoculum along with a mucilaginous substance, such as gastric mucin, which acts as an adjuvant, or assistant, of virulence. In a particular experiment, Griffith infected mice subcutaneously with nonvirulent R pneumococci derived from type I, but instead of using gastric mucin as an adjuvant, he mixed the inoculum with a thick suspension of S pneumococci of type II that had been killed by heat. He probably wanted to see if the killed S cells would facilitate infection by contributing to the system a heat-resistant "aggressin"—a hypothetical substance assumed by many bacteriologists to be responsible for virulence.

Whatever the exact reason that motivated Griffith to use heat-killed type II pneumococci as adjuvant material, the experiment was successful in the sense that the mice died of pneumococcal infection. However, the pneumococci that he recovered from the heart blood of these mice did not belong to type I, as he had expected, but to type II. Not only had the killed S pneumococci of type II helped the nonvirulent R cells to become virulent; in so doing, they had endowed the changed cells with their own capsular specificity. The R pneumococci derived from type I had been transformed into S pneumococci of type II. Furthermore, the pneumococci recovered from the infected mice continued to grow as S forms of type II when cultivated *in vitro,* even though no further type II material was added to the culture medium. The change was hereditary.

Griffith was so surprised by his unexpected findings that, according to a colleague who wrote his obituary notice in *The Lancet,* "he hesitated longer than most workers would have done before publishing these observations. He always took the line, 'Almighty God is in no hurry, why should I be?' "[35] Avery commonly took the same attitude, using almost the same expression, and, as we shall see later, he also waited several months before publishing the unexpected finding that the material responsible for type transformation is deoxyribonucleic acid.

Griffith had achieved the hereditary transformation of pneumococcal type by "accident," but he would not have recognized the accident or its implications if he had not been a keen observer and if he had not been extremely familiar with the behavior of pneumococci, both *in vitro* and *in vivo*. There is evidence, furthermore, that some of his earlier speculations had prepared him to accept the unexpected result. In the course of extensive epidemiological studies, which will not be reported here, he had been much impressed by the fact that, although pneumococcal types differ immunologically one from the other, there are great similarities among them. In his own words: "The various races of pneumococci resemble each other so closely in appearance of colonies and in the characteristic of bile solubility that *there can be no doubt that they belong to one species*" (italics mine).[36] He regarded all pneumococci as different breeds of a single species, which probably made it easier for him to believe that, under the proper conditions, they can produce one or the other material responsible for immunological specificity.

After having established that R cells derived from type I could be transformed into S cells of type II, Griffith suggested a Lamarckian explanation of the phenomenon. In his words, "the R pneumococcus in its ultimate form is the same, no matter from what Type it is derived; it possesses both Type I and Type II antigen in a rudimentary form or, as it may be differently expressed, it is able to develop either S form according to the material available."[37] He went so far as to suggest that the living R cells could use the components of the killed S cells "as a pabulum from which to build a similar antigen and thus to develop into an S strain of that type."[38] As is now known, this was a completely erroneous interpretation of the phenomenon. What happens in reality is that the gene corresponding to type II replaces the gene corresponding to type I in the cells undergoing transformation; these two genes exclude each other reciprocally, as if in competition for the same receptor. However, this mechanism could not be understood until the material responsible for type transformation had been isolated and shown to behave as a gene.

Griffith believed that the transformation phenomenon could explain obscure epidemiological aspects of lobar pneumonia, in particular the change of types that commonly occur from one outbreak to another: epidemiological problems were always his primary concern. Although he did try to reproduce the transformation phenomenon *in vitro*, he gave up after a few failures, probably because this did not appear to him to have crucial relevance to epidemiological problems.

Griffith's speculations on the epidemiological significance of the inter-convertibility of pneumococcal types remained virtually unnoticed, and have been largely forgotten, but his experimental findings had an immediate, enormous impact on immunologists all over the world. Within a few months after their publication, they were confirmed at the Robert Koch Institute in Berlin by Neufeld and Levinthal,[39] the same Neufeld who had first demonstrated the existence of different pneumococcus types 19 years earlier. Further confirmation came in 1929 from Hobart Reimann,[40] who was then working at the Peking Union Medical College, but was well known at The Rockefeller Institute, where he had served on the Hospital staff between 1923 and 1926.

Needless to say, Griffith's experiments were also widely discussed in Avery's department, but we did not even try to repeat them at first, as if we had been stunned and almost paralyzed intellectually by the shocking nature of the findings. Avery, in particular, found it impossible to believe that pneumococci could be made to change their immunological specificity. This reluctance was only natural on the part of a person who had devoted so much time, skill, and critical judgment to the doctrine of the fixity of immunological types. Furthermore, he was not the only one to be skeptical. Even in England, the Griffith phenomenon was not widely accepted; the 1933 edition of the immensely influential textbook on bacteriology and immunity by Topley and Wilson made only a hesitant mention of it in a short paragraph.

There were technical reasons for this skepticism. Griffith's method involved the subcutaneous injection into mice, along with the living R cells, of huge amounts of virulent pneumococci that had been heated at 60°C. Even though control tests seemed to prove that this temperature was high enough to kill all the virulent cells, Avery wondered whether a few might not have recovered their viability in the animal environment. These doubts were made more plausible because, in Griffith's own experiments, suspensions of S cells lost their ability to induce transformation when the temperature used to kill them was raised from 60°C to 80°C.

Although Avery had never met Griffith, or even corresponded with him, he greatly admired the latter's earlier scientific contributions, which were in fundamental agreement with his own views of pneumococcal biology. For example, Griffith's discovery that the colonies of virulent pneumococci are "smooth," whereas those of the nonvirulent forms are "rough," fitted well into Avery's scheme of virulence; colonial smoothness could be explained by the existence of a polysaccharide capsule around the cell. The change from R to S, described earlier by Griffith, was also

compatible with the concept of type specificity, because, until the 1928
study, reversion had resulted in the production of a culture immunologically identical with that of the S type from which the R form had originated.

The reversion of R to S of the same original type was indeed of such
interest to Avery that, in 1926, he had encouraged M. H. Dawson, a
young Canadian physician who had just joined his department, to investigate the conditions most favorable for the occurrence of the phenomenon.
Dawson first approached the problem by trying to determine whether all
the cells of a given R culture were capable of reverting to the S form. To
this end, he first prepared cultures derived from single R cells in order to
work with what he called "pure line strains"; he then subcultured these
clones in several different culture media. His results led him to conclude
that "the great majority if not all of avirulent R cells have the ability, under
the proper conditions, to revert to virulent, type specific capsulated organisms."[41] The phrase "proper conditions" meant that reversion was greatly
facilitated by the presence in the culture medium of growth-promoting
factors and of serum-containing antibodies to pneumococcus proteins.
Dawson's results, which confirmed and extended those of Griffith, thus
provided further evidence for the fundamental fixity of the specific types,
as they demonstrated that the attribute responsible for immunological
specificity and for type-specific virulence persisted in the R forms even
when it was not expressed.

The transformation problem was eventually taken up in Avery's laboratory by Dawson on his own initiative, simply because he believed almost *a
priori* that work done in the British Ministry of Health *had* to be right, and
that therefore Griffith's conclusions were valid. At first, he simply repeated Griffith's experiments, satisfying himself that pneumococci heated
at 60° and 70° could not multiply in mice, yet were capable of bringing
about type transformation *in vivo*. This initial part of his experimental
work was completed in late 1929 and published in 1930.[42]

Dawson then tried to carry out the transformation of types *in vitro*. In
collaboration with R. H. P. Sia, he cultivated R pneumococci in rich
culture media that contained antipneumococcus serum and heat-killed S
pneumococci. After several passages in such media, transformation occurred *in vitro*.[43] The success of this new experiment was due in large part
to the experience Dawson had gained in 1926 while repeating Griffith's
initial reversion experiments and studying the conditions under which R
organisms revert *in vitro* to S organisms of the original type.

In 1930, Dawson moved to the College of Physicians and Surgeons in

New York, where the pressure of clinical duties prevented him from continuing work on transformation. Avery, however, was now convinced that pneumococci could indeed be made to undergo transmissible changes in immunological specificity, and he encouraged J. L. Alloway, the young physician who had taken Dawson's place, to pursue the study of the phenomenon *in vitro*. Alloway soon demonstrated that transformation can be brought about, not only with whole, killed S cells, but also with a soluble fraction prepared from S pneumococci by "dissolving" the living cells in sodium deoxycholate, then passing the material through Berkefeld filters to remove cellular fragments. Furthermore, he found that the active material could be precipitated with alcohol, and thus obtained as "a thick syrupy precipitate" that was fairly stable.[44]

Alloway was an extremely shy person, little given to talk or to emotional expression, but he could not refrain from displaying his excitement when he noticed the extreme viscosity of active preparations containing the material that was then referred to in the laboratory as the "transforming principle." Alloway was thus the first person to handle the active, fibrous substance that was to be identified as DNA 10 years later. It should be mentioned in passing that, although Alloway's procedure is of great historical importance because it yielded the first soluble preparations of the transforming substance, it is extremely unreliable, because the preparations thus obtained are readily inactivated by the pneumococcal enzymes.

CHAPTER ELEVEN

HEREDITY AND DNA

The Transforming Substance and DNA

After Alloway's departure in 1932, Avery began to devote some of his own time to experiments on transformation, even though he was then actively involved in several other immunological studies. His initial goal was to improve the techniques for the preparation of the transforming substance and for the quantitative assay of its activity, but the work was full of frustrations. The irreproducibility of the transformation experiments during the 1930s was often punctuated with his remark, "Disappointment is my daily bread; but I thrive on it." When recalling this difficult period later, he was wont to say, "Many are the times we were ready to throw the whole thing out of the window." However, he did not give up, because he was served once more by his gentle stubbornness in the face of a problem of which he sensed the broad significance.

As on earlier occasions, he decided that the first essential task was to isolate the active material and determine its chemical nature. This does not mean that he shunned discussions concerning the mechanism of the transformation phenomenon or its bearing on other biological problems. The late Dr. James Murphy, for example, discussed with him some similarities between the transforming principle of pneumococci and the filterable agent that causes chicken sarcoma, referring to both of them as examples of "plasmagens." The relevance of type transformation to changes of a genetic nature, or to alterations caused by viruses, was frequently discussed in the laboratory. An echo of these discussions can be heard in the letter that Avery wrote his brother Roy early in 1943, at a time when much evidence was at hand that the transforming substance was a deoxyribonucleic acid (Appendix I). After describing in his letter what was then known of the material, he added playfully, "Sounds like a virus — maybe a gene. But with mechanisms I am not now concerned. One step at a time and the first step is what is the chemical nature of the transforming principle." He was excited by the fact that "the problem bristles with implications," but, with his usual discipline, he did not let speculations deter him from establishing the chemical identity of the active material.

Dawson had postulated that transformation was brought about by the capsular polysaccharide itself, acting as a template for its own replication. Later, Alloway suggested that the active material was a protein-polysaccharide complex that existed in the intact cell in the form of the complete capsular antigen. Avery considered all these possibilities, but some of his statements in the annual reports indicate that, as early as 1935, he had obtained the transforming material in a form essentially free of capsular polysaccharide and protein. Quoting from his own personal notes of 1936, Hotchkiss writes: "Avery outlined to me that the transforming agent could hardly be carbohydrate, did not match very well with protein and wistfully suggested that it might be a nucleic acid!"[1] However, the mention of nucleic acid at that stage of the work probably did not have much significance. Some of us were then working with a ribonucleic-acid component of the pneumococcal cell, and had found that it was attacked by the ribonuclease present in animal tissues.[2] It is therefore probable that Avery mentioned nucleic acid in passing, among many other possibilities that he had in mind. In any case, his fundamental strategy was to obtain the active material in a pure form, before attempting to determine its chemical nature.

Progress toward this goal was very slow for several years because of laboratory situations that will be discussed later and, in particular, because of two technical difficulties. Alloway's method for producing the transforming substance gave such erratic results that many preparations were essentially inactive. Furthermore, the assay method for determining activity was also undependable, and almost useless for evaluating quantitatively the concentration and purification procedures. These difficulties were resolved in the late 1930s and early 1940s when new methods were developed, especially through the work of Colin MacLeod, who had joined the department in 1935.

Techniques were developed for the production and centrifugation of very large volumes of pneumococcus cultures from which sizable amounts of transforming substance could be prepared. Instead of obtaining the material by dissolving the living pneumococci in deoxycholate, as Alloway had done, MacLeod first killed the organisms by heat and only then treated them with sodium deoxycholate. This procedure had the advantage of avoiding inactivation of the transforming substance by the pneumococcal enzymes, and thus gave more active and more stable preparations.

The discovery that the various R strains differ in the readiness with which they undergo transformation led to the selection of one particular strain which was particularly efficient in this regard, and which was used consistently from then on as a test organism in all the measurements of transforming activity. By several such technical improvements and by modifying the culture medium first devised by Dawson, MacLeod succeeded in working out a dependable assay dilution technique that permitted an approach to a quantitative evaluation of the various procedures used for concentrating and purifying the transforming substance.

These technical advances were not published at the time, nor were the initial discoveries to which they led concerning the properties of the transforming material. Yet, progress was extremely rapid, as can be seen from the facts described under the somewhat misleading title, "Studies on capsular synthesis by pneumococci," in the annual report to the Board of Scientific Directors for the year 1940–1941.[3] The report describes the preparation of the transforming substance by extracting heat-killed pneumococci with deoxycholate, precipitating the active material with alcohol, redissolving it in slightly alkaline salt solution, and shaking it with chloroform to remove proteins and other impurities. Preparations thus obtained appeared to be essentially "protein-free and lipid-free," yet they retained most of the original transforming activity.

The report stated also that "the transforming activity of purified extracts [is] resistant to the action of the crystalline proteolytic enzymes trypsin and chymotrypsin," as well as to the action of "a purified phosphatase prepared from swine kidney." Although the original extracts contained "considerable amounts of nucleic acid," this substance could be "almost completely removed by digestion with crystalline ribonuclease without affecting the transforming potency." There was no mention of deoxyribonucleic acid in the 1940–1941 report, but some of the observations prepared the ground for the later definition of this substance. MacLeod noticed that sodium fluoride protected the transforming material against the inactivating effect of pneumococcal enzymes, a fact he considered of interest because fluoride "is known to inhibit the action of esterases." Because animal serum and tissues rich in esterases also destroyed the transforming activity, there was some reason to believe "that the transforming principle may be an esterified compound"—a hypothesis formulated in the report for 1940–1941.

In the summer of 1941, MacLeod left The Rockefeller Institute to become Professor of Bacteriology at the New York University School of Medicine. He was replaced by Maclyn McCarty, a young pediatrician with extensive biochemical training. In some ways, McCarty's role in the identification of the transforming substance can be compared to that played by Michael Heidelberger 20 years before in the demonstration that the specific soluble substances of the pneumococcal capsules are complex polysaccharides.

Now that dependable methods had become available for the production of the transforming substance in a stable form and for its quantitative assay, McCarty could apply his chemical skills to the identification of the active material, under the guidance of Avery, with the help of MacLeod, who frequently returned to the Rockefeller laboratory, and with the encouragement of the Hospital director, Dr. Thomas Rivers. The outcome was that, within an incredibly short time, the transforming activity was shown to reside in a highly viscous fraction that consisted almost exclusively of polymerized deoxyribonucleic acid (DNA).

Surprising as it may seem, the Avery group never published a complete, detailed account of the steps that led them to this remarkable and unexpected conclusion. The facts are available in the laboratory records, but the pace of experimentation was so rapid during the last phase of the work that, according to Maclyn McCarty, who was then responsible for most of it, he himself now has difficulty in restructuring from his notes the intellectual processes that ended with the identification of the transforming substance as deoxyribonucleic acid. Since only he could tell the story, and should soon do it, I shall limit myself to the briefest summary of the different lines of evidence that began to accumulate after he started working with Avery.

One of McCarty's first experiments was to study the sedimentation of the active material in the high-speed analytical centrifuge, in collaboration with Alexandre Rothen. The results suggested that the molecular weight of the transforming substance is approximately 0.5–1 million.

All qualitative chemical tests on the sedimented material were essentially negative, except for those which suggested deoxyribonucleic acid. The results of quantitative elementary analysis, in particular the phosphorus-nitrogen ratio, also corresponded to deoxyribonucleic acid.

These findings were compatible with the results of earlier enzymatic studies and of new ones carried out by McCarty. Whereas, as has been mentioned, the activity was left unimpaired by treatment with a great variety of enzymes, including ribonuclease, the material was rapidly inactivated by all enzyme preparations capable of attacking authentic samples of deoxyribonucleic acids. As Avery was to write his brother Roy a few months later, the substance conformed "*very* closely to the theoretical values of pure *desoxyribose nucleic acid* (thymus type).* Who could have guessed it? This type of nucleic acid has not to my knowledge been recognized in pneumococcus before." Indeed, who could have guessed it?

---

* De*s*oxy- became deoxy- in the late 1950s or early 1960s. The decision to drop the *s* was made by an international nomenclature committee.

In early 1943, Avery, MacLeod, and McCarty presented all their findings to Drs. Jack Northrop and Wendell Stanley at the Princeton section of the Institute, but these experienced and famous chemists had no suggestions for further lines of evidence. Avery, however, still felt uncertain. When MacLeod asked him on the train ride back from Princeton, "Fess, what more do you want?" he could only shake his head; he wished for tests with a more purified preparation of deoxyribonuclease; this was to be satisfied by McCarty three years later. (Subsequently, Moses Kunitz of the Princeton group achieved the crystallization of this enzyme.)

The classic paper by Avery, MacLeod, and McCarty describing their extraordinary findings was submitted to the *Journal of Experimental Medicine* in November, 1943, and published in 1944.[4] However, all the basic experimental information had been accumulated much earlier, as shown by the mass of detailed evidence in the annual report submitted to the Board of Scientific Directors in early April, 1943.

It would be out of place to describe here the experiments that led to the identification of transforming activity with a DNA preparation, but it seems worthwhile to emphasize that the authors of the discovery owed much of their success to their skill in using a great diversity of techniques, ranging from the most physical to the most biological. The following outline of the preparative procedures described in the 1944 paper gives an idea of their technical ingenuity and boldness.

The pneumococci (type III) were grown in batches of 50 to 75 liters and separated by centrifugation in a steam-driven Sharples centrifuge. The cells were heated at 65°C to destroy the pneumococcal enzyme that was known to inactivate the transforming material. (In a later phase of the work, McCarty found that this enzyme requires magnesium and therefore is ineffective in the presence of citrate, which binds this metal. Addition of citrate to the medium thus resulted in much larger yields of the transforming substance.) The heated cells were washed repeatedly with salt solution, which removed large amounts of proteins, polysaccharides, and ribonucleic acid. The washed cells were then shaken with 0.5 percent deoxycholate, which extracted the active material. This material was separated from the deoxycholate solution by precipitation with alcohol. It was dissolved in salt solution and shaken with chloroform to remove most of the remaining proteins. Concentrated solutions of the active material were then treated with a series of enzymes capable of hydrolyzing proteins, ribonucleic acid, and the type III capsular polysaccharide; then they were once more shaken with chloroform to remove the last traces of proteins.

The final step was the repeated precipitation of the extract by the dropwise addition of one volume of absolute ethyl alcohol, with constant stirring. At this critical step, the active material separated in long, white, extremely fine "fibrous strands that wind themselves around the stirring rod." All those who have witnessed this delicate operation remember the excitement at the sight of the beautiful fibers, which were the purified forms of the viscous material that Alloway had first perceived 10 years before.

The fibrous character was due to the fact that the preparation consisted of highly polymerized deoxyribonucleic acid — DNA! Its biological activity was so great that it was capable of bringing about the transformation of R pneumococci into encapsulated type III pneumococci, even when used in a final dilution of less than 1 in 100,000,000 in the reacting system. The R pneumococci that had been transformed into type III pneumococci retained this newly acquired immunological specificity from generation to generation, even though no further transforming material was added to the culture medium. The transformation induced was therefore hereditary.

The statement that DNA was responsible for transforming activity had staggering implications. The effect observed was type-specific, so it followed that each type of pneumococcus had to have its own nucleic acid acting as a genetic bearer of immunological specificity; but this conclusion was incompatible with what was then taught concerning the chemical structure of deoxyribonucleic acid. P. A. Levene, the organic chemist of The Rockefeller Institute, was considered the world expert on the structure of this substance. He regarded DNA as a simple arrangement of nucleotides, and could not see how specific biological activity could reside in such a repetitious assemblage of phosphate, sugars, and nitrogen bases. He therefore concluded, as did many other chemists and biologists, that the specificity of the preparations was due to some contaminating substance.

Avery was haunted by the memory of the turmoil that had attended the announcement by him and Heidelberger, exactly 20 years earlier, that polysaccharides, and not proteins, were responsible for the immunological specificity of pneumococcal types. And he anticipated that even greater skepticism would now greet the claim of genetic specificity for deoxyribonucleic acid. For this reason, the manuscript of the paper reporting the claim was sent for publication only after it had been submitted for many months to the critical review and adverse criticism of associates and friends. Furthermore, the conclusions were presented with several cautionary statements. The authors recognized that it was "of course" possible that "the biological activity of the substance described is not an inherent property of the nucleic acid, but is due to minute amounts of some other substance adsorbed to it or so intimately associated with it as to escape detection."[5] This was a way of acknowledging the possibility that traces of some very active protein might account for transformation.

The mood of excitement tempered with caution that existed in 1943 in Avery's laboratory, and especially in his own mind, is well conveyed in his letter to his brother. Avery was then 65 years old and, having reached retirement age, had planned to join his brother's family in Nashville. However, the eagerness to complete his investigations on the transforming substance made him change his mind and stay in New York for a while longer. Although Avery ends his letter by characteristically apologizing for its lack of clarity and referring to it as "a rambling epistle," the document is, in fact, of considerable importance, because it was obviously composed with great care and presents many examples of the Professor's mannerisms (Appendix I). Here I shall mention only his emphasis on the need to document further the validity of the experimental findings, and his awareness of their large biological implications.

He emphasized that the thymus type of nucleic acids "were known to constitute the major part of the chromosomes but *have been thought to be alike* regardless of origin and species" (italics mine). This made it difficult to imagine how nucleic acids "*protein-free* could possibly be endowed with such biologically active and specific properties and this evidence we are now trying to get" (italics mine). After having described the work in detail, he asked Roy not to "shout it around" because "It's hazardous to go off half-cocked — and embarrassing to retract later. . . . It's lots of fun to blow bubbles — but it's wiser to prick them yourself before someone else tries to." (That he should be the one to prick his own bubble had long been an obsession with Avery.) In any case, the criticisms leveled against the claim that DNA was the transforming agent appeared to him sufficiently valid to warrant further studies.

Alfred E. Mirsky, who was working on biochemical genetics at The Rockefeller Institute, had pointed out that the evidence from enzymatic studies was not as convincing as was suggested in the 1944 paper.[6] Among his objections were that certain proteins are resistant to the proteolytic enzymes used by Avery and his colleagues, and that other proteins resist enzymatic action until they have been denatured. The illustrious geneticist H. J. Muller was so impressed by Mirsky's arguments that he wrote to the English geneticist C. D. Darlington in 1946: "Avery's so-called nucleic acid is probably nucleoprotein after all, with the protein too tightly bound to be detected by ordinary method . . . free chromosomes."[7]

Avery was, of course, aware of the criticisms leveled against the DNA theory, even though they were rarely expressed in public. That he resented them is indicated by the following statement in his 1946–7 annual report to the Board of Scientific Directors: "From the beginning *we ourselves have been keenly alert* to the possibility that the presence of some substance other than the desoxyribonucleate in our preparations may be responsible for the biological activity."[8] Even when, a few years later, Hotchkiss reported that the protein content of the most active preparations was below 0.02 percent, it remained possible, by interpreting this figure in the light of Avogadro's number, to postulate that the preparations contained a huge number of protein molecules, some of which might have biological activity. In fact, much of the work carried out in his department after 1943 was focused on the accumulation of additional evidence for the role of DNA in transformation.

Lest the phrase "his department" conjure a large and highly organized team of investigators, a few words should be said here concerning Avery's group during his last five years in New York. In addition to himself, the "team" consisted at any given time of only one or two scientifically trained persons, plus two technicians. As in the past, the unity of purpose in the laboratory emerged from casual conversations, rather than from formal organization.

McCarty was the only official member of the staff from 1943 to 1945; Harriett Taylor joined him in 1945, immediately after receiving her doctorate in genetics from Columbia University. McCarty left in 1946 to take charge of the rheumatic fever division in the Hospital, and elected not to continue working on the transformation problem because of the demanding nature of his clinical studies. Rollin Hotchkiss, who had long been interested in the transformation phenomenon, joined forces with Avery. He carried on after Avery left New York in 1948, whereas Harriett Taylor continued her work in France, where she moved in 1948 (she married the geneticist Boris Ephrussi). MacLeod, in his new professorial appointment at New York University, developed special biological aspects of transformation, enlisting as coworkers M. R. Krauss and R. Austrian. The contributions of this very small disseminated "group" confirmed the specific biological activity of DNA and enlarged the significance of the transformation phenomenon.

Briefly, the technique for assaying DNA activity was made simpler, and more reproducible, by two unrelated discoveries. It was found that serum albumin is required for transformation, and that the "competence" of R pneumococci to undergo transformation depends not only on the strain used in the test, but also on the phase of the growth cycle during which the organisms are exposed to the transforming material.

Tests with the purest preparations of crystalline deoxyribonuclease, obtained from the pancreas, showed that transforming preparations are inactivated by this enzyme, as it depolymerizes the nucleic acid, thus clinching the evidence for the essential role of the latter substance in transformation. Furthermore, a long series of elaborate chemical studies proved that proteins are not involved in the phenomenon.

Comparative analyses of DNA showed that the preparations derived from pneumococcus differ chemically from the classical thymus nucleic acid in having more thymine, less adenine, less cytosine, and a lower ultraviolet absorption per unit of phosphorus. By demonstrating that the molecular configuration of nucleic acids is not as rigidly programed as was once thought, these findings made it more plausible that DNA can exhibit biological specificity.

Transformation was extended to characteristics other than the capsular polysaccharide, for example to somatic components of the pneumococcus, to its fermentative activities, and to its resistance to various antibacterial agents. The latter properties lent themselves to the first truly quantitative assays measuring the numbers of cells actually transformed under varied conditions.

Genetic analysis of these phenomena indicated that transformation involves the transfer of chromosomal material to recipient bacteria, in which the material pairs with the homologous region of the recipient chromosome. Genetic linkage between different factors was also recognized in the transforming deoxyribonucleate agents.

I have listed chronologically, in references 9 to 42, 33 articles ranging from 1945 to 1960, by the four scientists—Colin MacLeod, Maclyn McCarty, Harriett (Ephrussi-) Taylor, and Rollin D. Hotchkiss—who were at one time or another directly associated with Avery in the transformation problem.

It is impossible to state precisely when the evidence became sufficient to convince the scientific public that DNA is involved in specific hereditary transformations. In fact, there was no way to obtain absolute proof by chemical techniques. Contamination of the most purified preparations by minute amounts of very active protein or other substance could not be ruled out entirely. Even the destruction of transforming activity by crystalline deoxyribonuclease was not foolproof evidence; despite its crystalline state, the nuclease might still have been contaminated with some other enzyme, as had been found for crystalline trypsin that turned out to contain some ribonuclease; in any case, the deoxyribonuclease might have acted on some hypothetical nucleoprotein involved in the transformation phenomenon.

It took an experiment, outside of the Institute, with a biological system completely different from that used by Avery to win universal acceptance for the genetic role of DNA. Using coliphage marked with ^{32}P (restricted to the DNA component of the virus) and with ^{35}S (restricted to the protein component), Hershey and Chase at the Cold Spring Harbor Laboratory showed in 1952 that most of the viral DNA penetrates the infected bacterium, whereas most of the protein remains outside. This finding suggested that DNA, and not protein, was responsible for the directed specific synthesis of bacteriophage in infected bacteria. In reality, the interpretation of this wonderful experiment was just as questionable on technical grounds as was the chemical interpretation of pneumococcal transformation, but its results were so completely in agreement with those obtained by Avery 10 years before, that the few remaining skeptics were convinced. The case for the view that DNA is the essential and sufficient substance capable of inducing genetic transformations in bacteria was not won by a single absolute demonstration, but by two independent lines of evidence.

Granting the importance and elegance of the Hershey-Chase experiment, the genetic role of DNA had become widely accepted before its results became known. In 1948, the year Avery left the Institute, an international congress was held in Paris, during which the problems of type transformation were presented by Rollin Hotchkiss and Harriett Taylor. At the end of the conference, André Lwoff of the Pasteur Institute interpreted the findings as follows:

> The study of the transforming principle of pneumococcus has led to the conclusion that the purine and pyrimidine bases are not present in equimolar proportions. This gives an inkling of a possible explanation for the specificity of nucleic acids. Once the transforming principle of pneumococcus is introduced into a bacterium it confers on it permanently a given specificity. But this principle is susceptible of modification and even at the present time we know of two varieties of specific nucleic acid of type III pneumococcus. They have been compared to allelomorphic genes.[43]

Thus, DNA had been incorporated into orthodox genetic theory five years before the Watson-Crick announcement of the DNA double-helical structure in 1953 started the deluge of work concerning the central role of DNA in genetic determinism.

During the 1940s and early 1950s, several lines of investigation that were seemingly unrelated to the transformation problem provided further indirect evidence for the role of DNA in the transmission of hereditary characteristics. Because these investigations were not carried out at The

Rockefeller Institute, it seems best to present them in an appendix, along with a few remarks concerning the large biological implications of Avery's original work on DNA (Appendix VI). As already mentioned, however, these implications were apparent to Avery and his colleagues long before the role of DNA in the transformation of pneumococcus types had been accepted as a landmark of biological history. Avery's very last annual. report to the Board of Scientific Directors was still conservatively entitled "Studies on transformation of pneumococci," and in it he affirmed once more that the objective of his work was to achieve a better definition of the chemical and biological factors involved in the phenomenon. However, this expressed only one aspect of his personality: his puritanical discipline as an investigator. Several years earlier, he had revealed another aspect when he had boldly and proudly announced to his brother and a few intimates that he and his colleagues had achieved, for the first time, a chemically directed modification of the genetic endowment.

Scientific Puritanism

Nothing was published on the transformation problem between Alloway's second paper in 1933 and the classic paper by Avery, MacLeod, and McCarty in 1944. This long period of silence has been interpreted by several authors as due to a failure by Avery to appreciate the full biological significance of the phenomenon, an interpretation seemingly justified by the fact that the 1944 paper includes only the barest mention of genes or viruses. Avery never made any public statement about the large biological implications of his findings, so it was assumed that, like Griffith, he was more interested in the immunological aspects of the transformation phenomenon than in its genetic aspects. Having been a member of Avery's department until July, 1941, and having remained in very close contact with him until he left New York, I know that this interpretation of his public silence is erroneous. From the time of Griffith's publication in 1928, the transformation phenomenon remained a constant topic of conversation in the laboratory. As to its significance for genetic theory and for viral biology, the only limits to speculation in the laboratory talk were the deficiencies in our collective knowledge of these fields and Avery's intellectual discipline, which kept unbridled intellectual free-wheeling under control.

I shall now present, on the basis of personal memories, the reasons for the *apparent* lack of progress during the decade between 1933 and 1943, and for the paucity of public statements regarding the great significance of the experimental findings.

During the 1930s, all members of the department were engaged in a large variety of investigations, most of which were extremely productive. Avery's own name appears on 25 papers between 1929 and 1941, and his actual participation in the laboratory work was considerable in all cases, although he was suffering from Graves' disease during the earlier part of that period. I can mention only briefly some of the fields of study that yielded results of theoretical and practical importance during that period: the chemical characterization of capsular polysaccharides; the synthesis and immunological study of artificial antigens; the autolytic processes in pneumococcus cultures and their bearing on the production of therapeutic sera; the recognition of the so-called C-reactive protein in the serum of patients during the acute phase of infection; the skin reactivity of animals and human beings to various components of the pneumococcus cell; the production and activities of a bacterial enzyme capable of hydrolyzing the type III capsular polysaccharide; the production and activities of the antibiotics gramicidin and tyrocidine (Appendix IV and Chronology II). To this list must be added the studies on streptococci that were carried out in the rheumatic fever department at the other end of the floor and in which Avery, as well as the rest of us, took a lively interest.

All the different projects in Avery's department were conducted in three small general laboratories and one chemical laboratory. All investigators participated in most of the experiments and in the interpretation of the results, regardless of their professional specialization and individual interests. As I try to evoke this period, I wonder how we could have found time to keep informed about the affairs of the outside world. But we did, especially when they had a bearing on our own scientific interests, as was obviously the case for Griffith's 1928 paper.

It could be argued that the slow pace of the work on transformation during the mid-1930s cannot be justified by the diversity of research projects, and that it suggests instead a failure to appreciate fully the importance of the phenomenon. In the words of one of my colleagues, "Generally when something as important as this is found, there is a concentration of effort to the exclusion of other avenues of research." The difficulty with this argument is that opinions concerning the importance of a problem depend upon the point of view from which that problem is considered.

Avery's department was in the Hospital; its staff was responsible for the care of respiratory diseases; its goal was the development of therapeutic sera and other procedures that offered hope for the control of lobar pneumonia. Much of the early work on transformation was carried out before the days of chemotherapy. Mortality rates were extremely high among patients suffering from lobar pneumonia, except in the case of type I, for which a therapeutic serum was available. By 1936, progress had been made toward the development of techniques for the production of sera effective against the other types. Under these conditions, I doubt that it would have been possible for anyone working in a hospital in direct contact with patients to concentrate all effort on transformation, and neglect the departmental commitment to the control of lobar pneumonia. What I find remarkable is that the problem was often given priority over the other urgent tasks.

(It is interesting to note in passing that the first important elaboration of the Avery findings on the genetic role of DNA was also due to a physician actively engaged in clinical work. Dr. Hattie Alexander was a pediatrician at the Babies Hospital of P&S when she demonstrated that strains of *Hemophilus influenzae* could be made to undergo hereditary changes in their immunological specificity by techniques similar to those used in the transformation of pneumococci.)

After the short initial period of hesitation, work on the transforming substance progressed rapidly, as shown by the fact that five papers on the topic were published between 1930 and 1933. Then the work slowed down because of technical difficulties. Even though transformation could be achieved *in vitro* with a cell-free material, the results were erratic. As has been pointed out, the methods developed by Dawson and Alloway were inadequate on two different grounds: the transforming material was unstable and the R strain used for the assay technique often lacked "competence" to undergo transformation. The countless experiments performed beween 1934 and 1940 to extend Alloway's findings did not lead at first to a systematic program, simply because the results were not reproducible.

In the late 1930s, the use of sulfapyridine and related drugs made it much easier to treat lobar pneumonia. This decreased the pressure for the development of therapeutic sera. Furthermore, World War II compelled the abandonment of certain research projects. As a consequence, it was easier for Avery and for MacLeod, who did not serve in the Armed Forces, to focus their thinking and efforts on the transformation problem. About that time, too, ways had been found to obtain more stable preparations and to select an R strain that was well suited to the assay technique. From then on, Avery and MacLeod, then McCarty after 1941, devoted much of their time to work on transformation, with the result that it took less than two years to isolate and characterize the active material.

In their 1944 paper, Avery, MacLeod, and McCarty expressed themselves in a very muted manner concerning the relevance of their work to genetics, but they were aware of its implications and "devoured genetical texts avidly." Avery's reticence was an expression of his intellectual self-discipline, which applied not only to modern genetics, but to much simpler biological problems, and was reflected in his scientific language. A few examples taken from daily conversations in the laboratory may help to establish the depth of his scientific puritanism.

He had doubts concerning the scientific validity of applying the Linnean system to bacteria, perhaps because the existence of a discrete nucleus in these organisms was not convincingly demonstrated until the 1940s. For that reason, he seldom, if ever, used the names *Diplococcus pneumoniae, Klebsiella pneumoniae, Mycobacterium tuberculosis;* pneumococci, Friedländer bacilli, and tubercle bacilli were good enough for him. For similar reasons, he did not use the classical jargon of genetics when discussing hereditary processes in bacteria. He spoke of transmissible properties, of bacterial dissociation from S to R, of reversion from the R form to the encapsulated form. It was certainly because the classical concepts of genetics had not yet been *proved* to be applicable to bacteria that, as late as 1948, he continued to use phrases such as "transformation of types" or "intraconvertibility of types" when referring to the phenomenon discovered by Griffith. As to the material responsible for the transformation of types, he was aware of its likely relation to a gene, but he felt more comfortable referring to it in cautious terms — first as the transforming principle, then as the transforming substance, and later as desoxyribonucleic acid, or DNA. In any case, he preferred the concrete meaning associated with the name of a substance to the ephemeral quality of such an abstract concept as gene.

On the other hand, he tried to learn as much as he could of classical genetics. From 1930 to 1948, "he collected, read, and commented on, with great interest and some amusement, the conjectures of many leading geneticists and biologists about transformation." In 1954, he turned these notes over to Hotchkiss[44]; they record, naturally, the reactions to Griffith's and his own paper, with suggestions by several geneticists that DNA was a chromosomal fragment acting a genetic role. He obviously had all these facts in mind when he stated in his letter to Roy, "By means of a known chemical substance it is possible to induce *predictable* and *hereditary* changes in cells. This is something that has long been the dream of geneticists. The mutation they induce by x-ray and ultra-violet are always unpredictable, random, and chance changes." But while he enjoyed such speculations, he considered it indecent to make them public if they went far beyond established facts. In the absence of really convincing evidence, it seemed to him merely clever, vainglorious, and indeed irresponsible to extrapolate from limited laboratory findings, however well documented, to sweeping statements that created false illusions in an impressionable public.

In his own subdued and smiling way, he showed signs of irritation when outsiders, whom he called "armchair biologists," explained glibly what he tried so hard to work out in the laboratory. I remember the pleasure he took at my French quotation of the Arab saying "Les chiens aboyent, la caravane passe" (The dogs bark, the caravan moves on), because it conveyed so well his deep feelings about the contrast between talkers and doers. But it is only since reading his annual report to the Board of Scientific Directors for 1946–1947 that I have come to realize the intensity of his irritation. There he clearly expressed in print, for the first and last time, his annoyance at those who assumed that he was not fully aware of the large biological implications of his own findings, and of the difficulties in ruling out all the possible sources of error:

> Various interpretations have been advanced as to the nature of this phenomenon. However, those of us actively engaged in the work have for the most part left matters of interpretation to others and have chosen rather to devote our time and thought to experimental analysis of the factors involved in the reaction. This is not to say that we are indifferent and have not among ourselves indulged in speculation and discussion of the relation of the problem to other similar phenomena in related fields of biology.[45]

He had enjoyed, as much as anyone, indulging in speculation and discussions concerning the relevance of his experimental work to other fields of biology, but only with associates and close friends — "among ourselves." Hotchkiss, who was his last collaborator, discussed the process of scientific discovery in words that would have been very congenial to The Professor. According to Hotchkiss, the first stage in making a discovery is one in which "faint evidence and speculation are encouraged," shared with associates and friends but "not the public"; in the second stage, the investigator should be overcritical, and communicate his findings in a way that will inform, but "not misinform or overinform . . . the people not fully able to evaluate the conclusions."[46]

Of course, the price of such thoroughness is some loss in the spectacular value of "discovery," and this was precisely the price Avery had to pay. His intellectual puritanism won him the admiration of those who were in direct contact with him, but it prevented him from gaining full recognition of his achievements by the outside world.

When Avery finally decided to retire in 1948, it had become clear that the method developed to isolate the DNA resonsible for the transformation of capsular polysaccharides could also be used to isolate other preparations of DNA capable of inducing other types of transformation. Moreover, great progress had been made toward defining the chemical and genetic aspects of the transformation phenomenon. In brief, the pneumococcus cell had been shown to contain a multiplicity of different forms of deoxyribonucleic acids, each endowed with a distinct biological specificity and located on chromosomes with functions similar to those of the classical chromosomes in the cells of higher organisms. It had become justified to equate the expression DNA with the word gene, both being abstract statements of the various chemical structures and genetic functions responsible for the specific distinctness of hereditary characters and for their transmission. Within less than 10 years after the type transformation of pneumococci by DNA had first been established — and regarded by many as a biological freak — the phenomenon had been incorporated into orthodox genetic doctrine. Furthermore, the discovery that DNA is the bearer of genetic information provided a chemical mechanism for genetic determinism, and thus created the new science of chemical genetics.

Most of the findings summarized in the preceding pages were published after Avery's retirement; his name does not appear as co-author of the publications in which they are described. However, much of the information is either reported or suggested in the annual reports that he continued to submit to the Board of Scientific Directors until 1948. The last phases of the experimental work carried out in his laboratory involved techniques that he did not have time to master, but the over-all program reflects his influence on the goals and general design of the experiments. In fact, he remained an actual participant in the laboratory work until he left the Institute, as reported by Rollin Hotchkiss:

> In his last two years at The Rockefeller Institute, Dr. Avery began his self-disciplined withdrawal from participation. At first he would disappear only when we (by that time only Harriett Ephrussi-Taylor and the writer) were planning experiments. I believe that he was determined not to be observed in any of the stages of ageing when he might be losing some of his mental faculties, as he had seen others do. This precaution was unjustifed, for his remarkable acuity and ability to focus never diminished. But the delight of performing experiments and observing the results he could not forego, and he would appear at the moment we commenced the work, asking "What are we doing today?" and start to help. We still enjoyed his influence at the time of discussing and interpreting the outcome. But this participation too he began to surrender,

especially in the last year, when I was attempting new chemical analyses, although all of his friends tried to make him welcome in the laboratories.[47]

A Premature Discovery?

Avery's work on DNA was published during a period of great excitement in theoretical biology — a period marked by new concepts of theoretical genetics and, in particular, by the flamboyant theoretical declarations of the "phage group."[48,49] Several schools of biologists, inspired by physicists who had moved into biology, made it fashionable to think about biological problems in terms of theoretical constructs, rather than of anatomical structures, physiological processes, and behavioral patterns; some biologists talked as if they were more concerned with cosmic riddles than with living organisms.

In contrast, Avery questioned the validity of biological generalizations and was even reluctant to use the word gene. He was virtually ignored by the theoreticians of genetics, precisely because he made no effort to communicate with them or, more exactly, to communicate to them what he had discovered by working at the bench instead of speculating about the secret of life. This peculiar scientific apartheid was still painfully evident as late as 1972, when Gunther S. Stent, an early member of the "phage group" published in *Scientific American* an essay entitled "Prematurity and Uniqueness in Scientific Discovery."[50] The theme of the essay is that "for many years" Avery's work on DNA "had little impact on genetics. The reason for the delay was not that Avery's work was unknown or mistrusted by geneticists but that it was premature . . . geneticists did not seem to be able to do much with it or build on it." The caption for the diagram that explained the experimental proof of the role of DNA ends with the phrase, "The significance of Avery's discovery was not appreciated by molecular geneticists until 1952," more than eight years after the details of the work had been made public.

As evidence for this extraordinary statement, Stent refers to the symposium, "Genetics in the 20th Century," held in 1950 to celebrate the golden jubilee of genetics.[51] He points out that, in the proceedings of the symposium, "Only one of the 26 essayists saw fit to make more than a passing reference to Avery's discovery, then six years old. He was a colleague of Avery's at The Rockefeller Institute, and he expressed some doubt that the active transforming principle was really pure DNA. The then leading philosopher of the gene, H. J. Muller of Indiana University, contributed an essay on the nature of the gene that mentions neither Avery nor DNA."[52]

This account of the published material is accurate in its essentials, but its interpretation appears in a different light when one knows—as Stent knew and should have mentioned—that the only member of The Rockefeller Institute staff present at the symposium was A. E. Mirsky, who still believed at that time that Avery's DNA preparations might contain small amounts of active protein. As to H. J. Muller, he had refrained from mentioning DNA in his formal lecture not for lack of awareness of its potential relevance to genetics, but because Mirsky's objections had made him uncertain concerning the chemical nature of the transforming substance. In the course of the general discussion, he did refer to a possible relationship between genes and DNA, but he added that "as yet no one has been able to correlate these features of chemical structure with the gene's peculiar property of self-reproduction." This was as positive a statement as was justified at the time, in view of the fact that the structure of DNA was still unknown.

The simplest way to discredit Stent's contention that "in its day, Avery's discovery had virtually no effect on *the general discourse of genetics*" (italics mine) is to quote verbatim a few of the many statements made by leading geneticists and theoretical biologists during the 1940s concerning the potential significance of the DNA work.

According to the account of Theodosius Dobzhansky, the eminent classical geneticist, he visited Avery's laboratory at least one year *before* the publication of the 1944 paper "and tried to argue that what were being observed were mutations like the mutations in Drosophila."[53] In the introduction to the second edition of his widely read book *Genetics and the Origin of Species,* dated March, 1941, Dobzhansky referred to pneumococcal transformation as follows: "We are dealing with authentic cases of induction of specific mutations by specific treatments—a feat which geneticists have vainly tried to accomplish in higher organisms."[54] Within a year after Avery's original publication, G. E. Hutchinson,[55] A. Marshak and A. C. Walker,[56] and Sewall Wright[57] suggested that DNA might be a chromosomal fraction acting a genetic role. G. W. Beadle was even more specific in his 1948 Silliman Lecture: "Pneumococcus transformations, which appear to be guided in specific ways by highly polymerized nucleic acids, may well represent the first success in transmuting genes in predetermined ways."[58]

Sir MacFarland Burnet visited Avery's laboratory in 1943 and immediately wrote his wife about the discovery, because he regarded it as "nothing less than the isolation of a pure gene in the form of desoxyribonucleic acid." And he said a few years later that "the discovery that DNA could transfer genetic information from one pneumococcus to another . . . heralded the opening of the field of molecular biology which has dominated scholarly thought in biology ever since."[59] In 1948, as already mentioned, André Lwoff had concluded the Paris Symposium on "Biological Units Endowed with Genetic Continuity"[60] with remarks expressing an attitude similar to that of Burnet. From these examples, it is clear that, contrary to Stent's assertion, the "general discourse of genetics" was immediately affected by the view that DNA is involved in genetic phenomena.

In his 1946 presidential address to The Royal Society, which included the citation of Avery for the Copley Medal, Sir Henry Dale stated that the transformation of pneumococcus type should be given "the status of a genetic variation; and the substance inducing it—*the gene in solution*, one is tempted to call it—appears to be nucleic acid of the desoxyribose type. Whatever it be, it is something which should be capable of *complete description in terms of structural chemistry*"[61] (italics mine). Here was a clear call to action, and it was answered at once by several chemists and biologists. At the Institute, Hotchkiss was beginning to study the comparative structure of deoxyribonucleic acids of different origins. Mirsky himself stated in 1947 that Avery's findings "have caused chemists to consider critically the evidence for uniformity among nucleic acids, and the generally accepted conclusion is that the available chemical evidence does not permit us to suppose that nucleic acids do not vary."[62] E. Chargaff stated emphatically that Avery's 1944 paper had been "the decisive influence" that led him to devote the major part of the activities in his department to the chemistry of nucleic acids.[63]

Awareness of the role of DNA in pneumococcal transformation led Hershey and Chase to design the ingenious experiment which established that, in bacteriophage infection, most of the viral DNA penetrates the infected bacterium, whereas the viral protein remains outside. Finally, it is certain that the findings of the Avery group were responsible for James D. Watson's decision to engage in the chemical program which culminated in the recognition of the double helix. According to Watson, his teacher, Salvatore Luria, had realized very early that "Avery's experiment made it [DNA] smell like the essential genetic material. So, working out DNA's chemical structure might be the essential step in learning how genes duplicated." When Watson arrived in England, he found that Francis Crick himself "knew that DNA was more important than proteins."[64]

In view of all these facts, it is obvious that the 1944 paper by Avery, MacLeod, and McCarty had a rapid and profound influence on both the thoughts and the laboratory programs of geneticists and other scientists. Apparently, certain members of the "phage group" regarded the orthodox chemical approach to the understanding of biological phenomena as pedestrian, too slow, and not revolutionary enough for their intellectual ambitions. They "did not seem to be able to do much with or build on it,"[65] because it did not fit their particular approach to genetics. It has been reported that Delbrück, the leader of the group, "deprecated biochemistry"[66] and even influenced some of his followers to avoid it. He "wanted to . . . go straight to the problems of gene replication and gene action"[67] and the "informational" approach seemed the most promising to this end. According to Stent, Delbrück and other members of the informational school even doubted that biological phenomena could be explained by the known laws of physics and chemistry; instead, they "were motivated by the fantastic and wholly unconventional notion that biology might make fundamental contributions to physics."[68] At least they hoped that information theory would give them rapidly, in the simplest possible way, some insight into the universal phenomena of life and especially into the mechanism of gene replication.

Avery's goal was less ambitious but more concrete. As in his earlier studies, he was interested in both the chemical composition and identity of the substances responsible for biological phenomena and in the mechanism through which they affect living processes. His years of experience in chemical immunology were his only contact with anything that might suggest an "informational" approach to biological phenomena, but he had not forgotten the lesson. He had assimilated Paul Ehrlich's classical (even though misleading, because oversimplified) pictures of antigens instructing the organs to produce antibodies that fitted the stimulating molecule as a piece of mosaic fits into a certain pattern, or as a key into a key hole. The immunological work done in Landsteiner's laboratory and in his own at The Rockefeller Institute transmuted this crude picture into a refined analysis of the specific relationship between the molecular structure of the antigen and the corresponding antibody. He was therefore not unprepared for the view that a nucleic acid of a certain molecular structure could instruct the pneumococcus cell to synthesize a polysaccharide endowed with immunological specificity. Hotchkiss has pointed out that, even though Avery was not a chemist and was unfamiliar with information theory, he nevertheless viewed biological problems in a very modern chemical light:

> . . . the specificity (now "information") was assumed to reside in individual molecular structures ("messages") capable of influencing (being

"translated") or interacting with (complex forming, repressing, etc.), cellular enzymes responsible for growth (biosynthetic systems). The confidence that a substance and an interaction underlie every manifestation[69] motivated his whole experimental approach.

The phage system, which the "informational" school of geneticists had selected because of the assumed fundamental simplicity of its replication mechanism, turned out to be far more complex than expected, almost as complex as a fruit fly. In contrast, Avery's down-to-earth chemical approach led, through DNA, to the formulation by Watson and Crick eight years later of the double-helix molecular structure[70] that provided the first material for effective thinking about biological information, thus making the dream of the "phage group" finally come true; but through the conventional channel of structural chemistry.

During the late 1930s, Avery had been nominated for the Nobel Prize in recognition of his immunochemical studies. After the 1944 paper, the Nobel committee was immediately alerted to the fact that he had once more made a fundamental contribution to biological science. But the 1944 paper was ineffective from the public relations point of view; it left open the possibility that some substance other than DNA might conceivably be involved in transformation; if failed to extrapolate from the role of DNA in a single bacterial species to the role of DNA in other living things. In other words, it did not make it obvious that the findings opened the door to a new era of biology. The Nobel committee, probably not accustomed to such restraint and self-criticism bordering on the neurotic, "found it desirable to wait until more became known about the mechanism involved in the transformation."[71] Yet, the very phenomenon of transformation, representing as it did the first example of directed change in hereditary characteristics, was in itself a biological landmark worthy of the Nobel Prize, regardless of the precise chemical nature of the transforming substance. But neither Fred Griffith nor Avery was a person who makes himself or his work obvious to international committees. They were not followers of fashionable scientific trends, nor did they attempt to create a fashion by broadcasting that they had reached a major turning point in the search for the secret of heredity. A day may come when the Nobel Foundation will review its errors of omission and write of Avery, as the Académie Française once wrote of Molière: "Rien ne manquait à sa gloire, il manquait à la nôtre."

CHAPTER TWELVE

AS I REMEMBER HIM

Gentle-Mannered and Tough-Minded

From 1927 to 1942, I worked in the laboratory adjacent to the one occupied by Avery at The Rockefeller Institute. He never closed the door of his own laboratory and but rarely that of the small office attached to it, so that I was witness to most of his activities at the bench or in conversation, and also to his interludes of day-dreaming. As I was a rather quiet person, he was hardly aware of my presence, and often behaved as if he were alone. This gave me a chance to observe certain aspects of his personality quite different from those that appeared on the façade he exposed to the public under the normal conditions of everyday life. Some of the moods he displayed puzzled me at the time, but I understand them better now that I realize how much his quality as a person depended upon the constant discipline he exerted over himself.

The appellation "Fess" brings to mind, first and foremost, a slender man always dressed in a neat and subdued style; his conservative appearance added to the charm of his lively and affable behavior. The most dominant features of his physical being were his sparkling and questioning eyes surmounted by the bulky dome of a head that appeared too voluminous for the frail body. In repose, his face expressed a gentle, quiet wisdom, which was enlivened by a warm smile of welcome and by effusive greetings as he met colleagues, friends, or strangers. He transformed even the most casual conversation into an artistic performance by a panoply of words and gestures that managed to be simultaneously spirited and restrained. Each of his statements was spiced with mimicry, pithy remarks, verbal pyrotechnics, and picturesque analysis. The extroverted playfulness of his nature, and his phenomenal empathy for every person or situation that engaged his interest, made any contact with him an intellectually rewarding experience, always entertaining and often enchanting.

Now and then, however, he displayed other patterns of behavior that seemed rather disconcerting and at first sight less attractive, yet had a haunting quality. On the rare occasions when he was alone, he was prone to move slowly from one object to another in his laboratory, gently whistling to himself the lonely tune of the shepherd's song from *Tristan und*

Isolde. His gaze was then focused inward, and his brooding forehead appeared almost to dwarf his body. For a few fleeting moments, he seemed to be a melancholy figure out of contact with the external world, but this attitude of inwardness vanished as soon as an occurrence brought him back to reality.

If a colleague or a visitor walked into his room during one of these moods of withdrawal, he would immediately be welcomed with Avery's usual warm smile. If the telephone rang, commonly for an invitation to dinner or to some other social event, his response was instantaneously one of joyful thanks or of profuse regrets, couched in endearing or apologetic terms. These interruptions, however, did not really break the spell of his inward mood. Once the visitor had walked out of his office or the telephone conversation had ended, an expression of lassitude was likely to reappear on his face, as if a smiling mask had been removed. He would push the telephone away from him in an abrupt gesture that suggested irritation against encroachment into his privacy. His smile was replaced by a tortured expression of protest against the need to play a social role that he resented because it did not fit his present mood. He certainly suffered from his own attitude on these occasions, because, as he was wont to say, obviously referring to himself, resentment hurts the person who resents, much more than the person who is resented.

While he never engaged in criticism or unkind gossip, he manifested his feelings of censure in other ways. What he did not approve, he simply ignored. His eagerness to avoid certain social roles probably accounted for some of his behavioral peculiarities. For example, he left many letters unanswered and did not want to have a secretary, even though the large number of scientists in his department would have justified one during the 1930s. The departmental manuscripts and administrative matters were handled in the office of the Hospital director. He refrained from reviewing or sponsoring scientific papers unless he had had a direct part in the performance of the experiments; I can still see him graciously but firmly pushing back under the arm of a visiting bacteriologist, whom he had courteously entertained for more than an hour, the manuscript of a paper that the visitor wanted him to endorse for a certain journal. He was extremely selective in what he gave of himself, even to his colleagues; for example, he would act as if he had not noticed the presence of one of his young associates who tried to approach him, but whose attitude he found distasteful or simply irritating.

Thus, while he was exquisitely gentle-mannered, he was tough-minded; in his own subdued way, he was indeed ruthless with regard to what he elected to do or not to do. Once he had decided on a course of action, he did not allow any external influence to deter him from reaching his goal or to force him into an activity he did not desire.

These different views of Avery's behavior — his extroverted attitude and his inwardness, his graciousness and his toughness of spirit — are not as incompatible as they appear on first sight, but rather correspond to complementary aspects of his nature. His effusive welcome, his receptiveness and responsiveness to new situations and to new persons, expressed his eagerness to perceive all aspects of the external world. These qualities accounted for his ability to identify himself with new scientific problems or new ways of life. In contrast, most of his introverted moods probably occurred in the periods during which the impressions he had received and the phenomena he had observed became integrated with his own substance in the patterns that formed his self-created persona. He had been endowed by nature with many intellectual gifts, great sensitivity, and an immense skill in dealing with people, and could thus have been successful in many different types of activities and environments. Indeed, one of the most interesting aspects of his life is that each period of it provided him with the chance to give successful expression to one or another facet of his rich personality.

Avery's Consecutive Persona

Avery's neighbors during his retirement years in Nashville must have wondered why such a kind and attractive gentleman had remained a bachelor. In fact, there were other mysteries in his life. After having majored in humanistic subjects at Colgate University until the age of 23, how did he manage, within a few years, to redirect all his energy and talent to the study of biomedical problems? Since he emphasized declamation and debate while at college, and had been highly successful as a teacher during his early medical days, why did he seem to resent lecturing on his own research after he became a famous scientist? He was an extrovert in youth, as shown by his eagerness to play the cornet in front of the Mariners' Temple, by his position as leader of the college band, by his participation in college debates and his dramatic proclamation concerning the existence of God on the steps of the Colgate Alumni Hall. What circumstances made him an introvert and shun public appearances during his mature years? Avery did not discuss the reasons for these extraordinary mutations of his persona, even with his closest associates. The mysteries of his behavior call to mind Dr. William Henry Welch, whose inner life also remained a closed book, even to those who knew him best and whom he regarded as his trusted friends.

While a student at Yale, Welch had desired to become a professor of

Greek, but when the opportunity came to him to fulfill this wish, he elected instead to go into science. He enjoyed social contacts and was extremely popular with women, as well as men, but he never married and had very few, if any, really intimate friends. He came to be known all over the world as the congenial and jovial Popsy, but no one had access to his private world. Dr. Simon Flexner, who had been closely associated with Welch for some 50 years, stated in the biography he wrote in collaboration with Thomas Flexner that aloofness was at least as much a characteristic of Welch as was his congeniality, and that he never allowed social relationships to intrude into his privacy or overcome his emotional reserve. Welch's life was governed by this aloofness until the very end, as shown by the account of his death from cancer in the Flexner biography:

> Welch was holding to his lifelong habit of not confiding in anyone, irrespective of who that person was or what he knew. Always he had been surrounded with people, and during most of his life he had moved on a public stage toward public ends, but always *he had kept the inner core of his being inviolate.* And when the final trial came, he did not change. While his body suffered, his mind struggled to maintain before the world the same placid exterior that had been his banner and his shield. Popsy, the physician who had been so greatly beloved, died as he had lived, *keeping his own counsel, essentially alone*[1] (italics mine).

Judging from Avery's behavior in difficult periods of his life, it is certain that he, too, wanted to keep his own counsel and face his destiny alone. He did not discuss his health when he was suffering from Graves' disease, except to answer his friends' questions with the statement that he was feeling much better. He never mentioned concern for members of his family, even though their problems were much on his mind. He did not express irritation at criticisms of his work, even when these were unjustified. Since he left no record of his personal thoughts, I shall attempt to imagine, from very tenuous clues, some of the factors that may have influenced important decisions of his life. One justification for this questionable exercise is that it will provide an opportunity to bring out a few more aspects of his many-faceted and endearing personality.

On an early spring day in 1934, I informed him that I was about to get married. He immediately rejoiced at the news, and described with animation how this change would enrich my life. At one point in our conversation, he slowly walked to the window and looked outside, lost in thought for a few seconds. Coming back to his chair, he casually mentioned that he, too, had contemplated such a move years before, but that circumstances had stood in the way of his plans. Then he turned the conversation back to my own life, although his attitude tacitly expressed a longing for the kind of intimate companionship which he had not known. One of the great joys of life, he remarked in passing, is to go home to someone who would rather see you than anybody else.

I have recently been told that, while talking to the wife of one of our colleagues, he referred on several occasions to a certain nurse who had meant a great deal to him. To this information, I can add that he took special pleasure in mentioning the course he gave to student nurses at the Hoagland Laboratory, at the time when the success of his lectures won him the appellation "The Professor." He was then 32 or 33 years old, and it is not unreasonable to imagine that he developed an emotional attachment to one of these young women.

His years at the Hoagland Laboratory, however, must have been an anxious period of his life. He was training himself for laboratory research and, although he was involved in several scientific problems, his professional future was still uncertain. He felt responsible both for his young brother Roy and for his orphaned first cousin Minnie Wandell, whom he supported for the rest of his life. Throughout the years, his correspondence with his brother Roy and his sister-in-law Catherine leaves no doubt that he was willing to sacrifice his personal desires to the welfare of his family. One of the reasons he remained a bachelor may therefore have been that he thought this was the only way he could properly fulfill the familial obligations he had inherited. It is also probable that he eventually found it increasingly distasteful, and even painful, to accept any commitment, except those he took for granted as head of his family, that would impinge on his intellectual and emotional freedom.

Avery always retained a profound sense of responsibility toward others, but he displayed throughout life a remarkable ability to change his persona according to circumstances and to the roles he elected to play.

He had joined the Baptist church at the unusually young age of eight years and had taken an active part in its activities at the Mariners' Temple. His father was dead by the time he entered college, but his mother was then heavily involved in the affairs of the Baptist Mission Society. It would therefore have been natural for him to train for the ministry, as did many of his classmates at Colgate. However, his religious beliefs evolved during his early college years, and he probably found it impossible, or at least intellectually dishonest, to follow in his father's footsteps. A career in medicine may have seemed to him a proper substitute for the Baptist ministry.

From his own accounts, he was soon disappointed by medical practice.

By necessity, the clinician and the public health officer must act even when they do not understand the inevitable complexity of the disease process with which they are dealing. For lack of sufficient information, they must commonly make value judgements on the relative importance of the multiple factors that impinge simultaneously on the patient or population group for which they are responsible. This situation must have been particularly difficult for Avery, who practiced medicine during the early 1900s, at a time when an educated physician had enough general scientific knowledge to be aware that few of his interventions were scientifically justifiable and usually were, at best, completely empirical.

As he worshipped rational thought, he probably turned to laboratory research both because this offered the best approach to progress in medical practice, and because it provided him with the opportunity to deal with experimental situations that he could understand and control. Whatever the state of knowledge, the experimenter has the freedom to separate from the complexity of natural phenomena a few limited aspects that he chooses to investigate; instead of dealing directly with the confusing complexities of natural processes as he encounters them in the raw, he can often create experimental models simple enough to be controlled and manipulated at will — although at the risk of working with artificial situations far removed from reality. As already mentioned, Avery enjoyed *observing* phenomena in natural situations, but when it came to the systematic study of them, and particularly to active intervention into them, he suffered acutely, and was almost paralyzed until he had reduced the complexity of the system so as to control its variables. Once he had made this choice, he settled into the way of life of a laboratory scientist, and never departed from it until the time of his retirement.

Another aspect of his personality may have been influential in his decision to shift from clinical medicine to laboratory research. Because there were few really effective therapies in the early 1900s, taking care of the sick largely meant providing them with psychological comfort. In the words of Francis Peabody, the most important aspect of the care of the patient is caring for the patient. Avery was certainly capable of the human understanding and sympathy implied in Peabody's phrase; this compassionate approach to medicine, however, is emotionally demanding of the physician if he identifies himself completely with his patients. Indeed, many physicians experience difficulties on this score during their early clinical experience, but most learn by practice to display kindness without becoming emotionally involved; they develop the skill to turn emotional involvement off and on as needed. It is possible that Avery's temperament made it difficult for him to achieve this protective kind of behavior. The facts that, despite his great sensitiveness and his ability to inspire affection in all the persons with whom he came into contact, he never married and had very few really intimate friends, suggest that he tried to avoid deep emotional commitments. He may have found it painful to achieve the right balance of involvement and detachment that is essential in clinical practice.

Avery always had a whimsical smile when, in the course of laboratory conversations, his young associates made dogmatic statements about such nonscientific topics as social problems, the management of institutions, the characteristics or activities of important persons. I can still hear the gentle irony in his voice when he asked us on such occasions, "Now, are you *really* sure of that?" He was probably the more amused by our cockiness because he remembered that he, too, had often been guilty of unwarranted statements during his youth and early adulthood. In college, as already mentioned, he had engaged in brash talk on almost any subject. At The Rockefeller Institute, he had published in 1916 and 1917 hasty conclusions that were soon proved to be erroneous (Chapters Seven and Eight). He therefore knew from experience the human propensity to use facts solely for the sake of rhetorical effects and to ignore facts when they stand in the way of one's prejudices.

Self-knowledge had made him wise, but he understood that wisdom, far from being an innate attribute, must be constantly gained by the mistrust of spontaneous impulses and by self-mastery. He was always immensely successful on the few occasions when he accepted invitations to give public lectures, so it is unlikely that fear of public reaction made him try to avoid this kind of activity during his late professional life. It is more probable that he was afraid of himself and, especially, of exerting an influence that did not correspond to what he considered as really significant truth. There may have been an element of conceit in this attitude, since it implied that anything he said was likely to receive public attention and to have public effects — as indeed was the case. This conceit was an expression of the self-confidence that his schoolmates had noted in the Colgate yearbook and that was now muted by mature wisdom.

On the other hand, Avery may have refrained from public speaking for the same fundamental reason that made him remain a bachelor, abandon humanistic studies and other interests from which he could have derived satisfaction. Increasingly, he refused to consider any commitment, to either a person or a cause, that would interfere with the few roles he had accepted or elected to play. This refusal even extended to honors that he greatly valued. In 1944, he was proposed for an honorary degree by Cambridge University; in 1945, he was awarded the Copley Medal by the Royal Society of London. Even though he was a great admirer of British culture, he refused to go to England on both occasions, saying that his state of health did not permit him to travel except by first class, and that he could not afford the expense. This was, of course, a lame excuse, because foundations would have been willing to finance his trip.

As he did not go to England, he could not receive the honorary degree from Cambridge University, but Sir Henry Dale, who was then President of the Royal Society, decided to bring him the Copley Medal at The Rockefeller Institute. Dale was accompanied by Dr. Edgar Todd, who knew Avery and who has told the story of the occasion. The two English visitors arrived at the Institute unannounced, and went directly to Avery's department, with which Dr. Todd was familiar. As they approached, they saw Avery, alone in his laboratory, manipulating pipettes and test tubes and transferring bacterial cultures. As they retreated without letting their presence be known, Sir Henry Dale said simply to Dr. Todd, "Now I understand everything." [2] What he had understood was that Avery had elected to be a laboratory scientist and that he resented being distracted from his self-appointed task.

The roles selected by Avery naturally changed in the course of his life, except for his familial responsibilities, which are repeatedly mentioned as taken for granted in his correspondence. What remained constant was his determination to be what he wanted to be, at any given time. This spirit of determination can be read in all photographs of him, even those taken when he was very young. It found expression in the several consecutive social roles that he played — as the child intensely involved in the activities of the Mariners' Temple; as a humanist and public figure at Colgate University; as the scientific investigator who emerged at the Hoagland Laboratory and came to flowering at The Rockefeller Institute; finally as the country gentleman, who delighted his family and his neighbors during his retirement years in Nashville.

Both his singleness of purpose and his ability to adapt to new circumstances are obvious throughout the 35 years he spent at the Institute. His scientific achievements during that period naturally brought him many offers from other institutions, but he ignored them. The Institute provided for him an ideal environment, because, although he was responsible for a well-defined professional task, he had otherwise full freedom to develop the intellectual schemes he nursed in his mind.

An Unspoken Scientific Philosophy

Although Avery had received extensive training in philosophy at Colgate, he shunned philosophical discussions about science and scientists.

The nearest he came to a formulation of his views about the scientific method or the social obligations of scientists was in the 1941 speech he delivered when he was president of the Society of American Bacteriologists.[3] The speech was well-received, yet he refused to publish it or even to deposit it in the archives of the Society of American Bacteriologists, as is the usual practice. I retrieved a copy of it from a waste basket where he had discarded it in 1948 while clearing his desk before retiring to Nashville. The speech is of interest, not for the originality of the ideas it presents, but for what it reveals of Avery's mannerisms and attitudes.

The typescript I recovered is fortunately the one from which he spoke. It shows several modifications of the typewritten text in his own handwriting and also numerous indications in pencil as to nuances of expression, almost like musical notations (Appendix II). Looking at the script, I can hear the points of emphasis and of query, the linkages between words and phrases, and inflections of voice that enabled Avery to convert the somewhat artificial and labored text into an exciting and seemingly spontaneous performance.

The theme of the second half of the speech, which is of little interest because it is conventional and stereotyped, deals in abstract terms with the comparative importance of theoretical and practical science, and with the moral and social obligations of scientists. Most of the views expressed in this section are quoted from Sir Robert Gregory, who had just retired as President of the British Association for the Advancement of Science, and more extensively from Raymond B. Fosdick, who was then President of The Rockefeller Foundation and the brother of Avery's classmate at Colgate. The quotations were all of unobjectionable character, as could be expected from such orthodox officials.

Avery's own statements about science and scientists are just as conventional as those he quotes. According to him, scientists in general, and microbiologists in particular, "have undeniably and always been in the service of human welfare. . . . It is the ancient tradition of the spirit of science that it follows no flag, recognizes no geographical boundaries, and sets up no trade barriers. Complete freedom of scientific thought, and the free interchange of knowledge are prerequisites for the survival of the spirit of free inquiry. They are to the Commonwealth of Science what the Bill of Rights is to the life of democracy."

Avery believed, of course, in the ideals he thus set forth, but it is obvious from his abstract formulation of them, so different from the sparkle of his usual manner of speech, that he was not much interested in the topic and, in any case, had no desire to challenge conventional social values. His behavior on the lecture platform did not differ in this respect from what it was in the laboratory and in private. Whatever thoughts he had about people and institutions he kept to himself. He gave the impression that he had decided irrevocably, at some time in the past, not to attempt to change the world of men (except himself) and instead to focus his attention on understanding other forms of life, the smaller the better.

Judging from the notations for emphasis on the typescript of the speech, the statements made by Fosdick that were most meaningful to Avery were those concerning science as a way of thinking about the world:

> Science is more than the technologies that cluster about it — more than its inventions and gadgets. It is even more than the discovery and correlation of new facts. Science is a method, a confidence and a faith. . . . It is a confidence that truth is discoverable. It is a faith that truth is worth discovering

Avery may have doubted that all forms of truth are really worth discovering, but the belief that the scientific study of medicine can lead to the discovery of large biological truths was the faith by which he functioned in the laboratory. It was therefore proper for him to conclude with the statement that "science, in obeying the law of humanity, will always labor to enlarge the frontiers of knowledge" — an irreproachable platitude he had borrowed from Pasteur.

He was truer to himself in the first half of the speech, where he expressed some of his characteristic attitudes as a laboratory worker. He wanted to convey the view that no one should regard his "own corner of knowledge as the source and directive of all biological thought." Instead of stating this truth in abstract academic sentences, however, he took an obvious pleasure in quoting the words that Fosdick had used to make fun of those scientists who believe in the unique importance of their own discipline:

> Choose off the shelves a group of learned treatises and sample the prefaces: *Mathematics:* — it is the queen of sciences; *Physics:* — it is the source of the basic laws for the behavior of all matter; *Chemistry:* — a recent text says, "Chemistry touches all human interests. It is the central science"; *Biology:* — it assaults the greatest mystery of all, the mystery of life; *Astronomy:* — it has the cosmos and eternity for its heroic theme; *Philosophy:* — it is an examination of the ultimate questions which give life meaning. And so one could expand the list, with brave and startling claims for the central character and basic importance of one field, one specialty, one segment of knowledge after another.

Avery played with the thought that a learned treatise on microbiology might justify itself with the statement, *"Microbiology:* — It is the king of sciences, it assaults the citadel of life's deepest mystery, the microcosm — the potentialities of which challenge the human intellect." But while he spoke with tongue in cheek of certain excessive claims made for microbiological sciences by some of his colleagues, later in his text he could not refrain from quoting example after example illustrating how microbiological studies have, in fact, illuminated a wide range of biological problems. Addressing physiologists and biochemists, he admonished them jokingly, "Go to the microbe, thou scientist, consider its ways and be wise." This paraphrase of the Biblical saying was obviously meant to entertain his listeners, but it expressed also his scientific philosophy as a biologist. He believed in the interdependence of all natural sciences; he regarded chemistry as playing an essential role in biological progress; he was convinced that the chemical unity of life could best be documented through the study of microorganisms.

Avery always refrained from extending scientific concepts into domains where they could not be converted into experimental laboratory tests. In particular, he avoided scientific discussions of a philosophical character about the human condition or the origin of life. This restraint was not due to lack of interest in these problems, or to ignorance of them. In his early years at The Rockefeller Institute, he had been exposed to Jacques Loeb's assertions that free will and ethical attitudes can be explained by physicochemical mechanisms. In the 1930s, he had been witness to free-wheeling discussions about the origin of life, generated by the finding that the tobacco mosaic virus can be obtained in a crystalline form. At the time he delivered his presidential speech before the Society of American Bacteriologists, he was in the process of demonstrating that DNA is the bearer of hereditary characteristics in pneumococci, and he certainly realized that this discovery would lead to speculations about the "nature" of life. In other words, he was fully aware of the general tendency to read philosophical implications into any new form of biological knowledge. However, the way he shook his head when such discussions went on around him made it clear that he did not have as grandiose and sweeping a view of these implications as did some of his colleagues. In my judgment, he felt that certain aspects of life and certain areas of human concern are outside the domain of science because they cannot be formulated in such a manner as to be put to the test of verification or falsification. Because he did not explicitly state his opinions on such matters, I shall try to imagine them, as much from his silences as from fragmentary statements he made now and then concerning the limitations and potentialities of the scientific approach.

To begin with, one can take it for granted that, if he had elected to discuss problems of scientific philosophy, it would not have been in abstract terms, but through illustrative examples. For example, he might have said that the phrase "God exists" is a statement which has meaning for those who make it, but is not scientific because there is no conceivable way that it can be proved right or wrong. Similarly, when Gauguin inscribed on his famous Tahiti painting "Where have we come from? What are we? Where are we going?" he was asking questions which are of universal significance, but which are not answerable in scientific terms at the present time . . . if ever. In fact, there is probably no way to give scientific answers to such questions as What is the nature or purpose of the universe? Of life? Of consciousness? Of free will? These are truly metaphysical, in the Greek sense of the word.

While Avery never discussed such questions, he did believe that scientists can provide knowledge relevant to them by converting them into other questions amenable to experimental tests. For example, scientists cannot usefully discuss the *nature* of the universe, but they can make testable and falsifiable statements about its components and its development; they cannot discuss the *nature* of life, but they can investigate the mechanisms of growth, self-reproduction, and evolution in living things; they cannot discuss the *nature* of free will, but they can determine the influence of prior conditioning, of the state of health, and of the total environment on the ability of human beings to make choices and decisions.

Avery also shunned theoretical discussions about the scientific method. If he had been familiar with Karl Popper's writings, he would have agreed with him that the method involves a number of different consecutive steps, such as the recognition of a problem; imagining solutions to it in the form of hypotheses; deducing testable propositions from these hypotheses; trying to confirm and refute the hypotheses by experiments and arguments; selecting among competing theories. But he would have suggested gently that effective scientists intuitively go through these steps without bothering to formulate them in philosophical terms.

His own way of acknowledging the existence of a philosophy of the scientific method was to indoctrinate his young associates with picturesque admonitions. For example, he would welcome any failure or inconsistency in experimental results with the remark, "Whenever you fall, pick up something." When in search of an explanation, he would assert, "Be fearless when it comes to hypotheses, but humble in the presence of facts." As to Karl Popper's famous law that falsifiability is the criterion of demarcation between science and nonscience,[4] he would have stated it in the form of his favorite phrase, "It is great fun to blow bubbles, but you must be the first one to try to prick them." Without taking the trouble to say it, he would also have agreed with Sir Peter Medawar that science is "the Art of the Soluble," [5] and he would have especially emphasized that good scientists have the wisdom not to deal with problems that lie beyond their competence or outside the domain of science; they intuitively elect to study the most important problems they can solve.

On the other hand, solving problems simply because they can be solved did not seem to him a reasonable occupation. It was the kind of activity that he described with a smile as "pouring something from one test tube into another." His attitude toward busy-ness in science was much the same as that expressed by Sir Joshua Reynolds about painting: "A provision of endless apparatus, a bustle of infinite inquiry . . . employed to evade and shuffle off real labor . . . the real labor of thinking." Avery, as mentioned earlier, symbolized the very opposite of this attitude. He spent countless hours debating what was really important among the countless things that could be done, and once he had made his choice he moved toward his goal with great economy of effort and material.

Originality and Creativity

Watching Avery at work in the laboratory was an unforgettable experience. Because he abhorred complex situations and a plethora of equipment, his own presence was the essential part of the show. His gaze was intensely focused on the operation being carried out or on the phenomenon under observation; his movements were limited, but of extreme precision and elegance; his whole being appeared to be identified with the sharply defined aspect of reality that he was studying. Confusion seemed to vanish wherever he functioned, perhaps simply because everything became organized around his person.

His attitude in the course of an experiment had many similarities with that of the hunter in search of his prey. For the hunter, all the components — the rocks, the vegetation, the sky — are fraught with information and meanings that enable him to become part of the intimate world of his prey. Just as the hunter penetrates that particular world, so did Avery penetrate the world of the phenomena he studied. He invested his whole attention so completely in the problem at hand that he became inattentive to extraneous matters. When at work in the laboratory, he found it difficult to concern himself with the questions asked of him unless he could relate them to his own problem.

Like the hunter, also, he took more pleasure in the pursuit of the prey than in the outcome of the hunt. The solution of a problem brought him only transient satisfaction; he was chiefly attracted by the unknown, and found charm in established knowledge chiefly to the extent that it helped him in his own research. Thus, he continued to stalk new phenomena for the sake of the hunt itself. One could have applied to him Pascal's paradoxical saying that he was not so much in search of truth as in search of the search for truth.

Since the search as a process, rather than the product of the search, was the more appealing aspect of scientific work for Avery, he could honestly say, as he was prone to do, that he would have been just as interested working with the hay bacillus as with the pneumococcus. In fact, many of the problems on which he worked were not of his own choice. He did not imagine them; at the most, he selected among those provided by the conditions of his time and of his milieu. To a large extent, indeed, his scientific problems were almost forced on him by his social environment.

When he began working at the Hoagland Laboratory, for example, acidified milks of the yogurt type had just become popular and were important commercial products in his Syrian neighborhood; he therefore studied the lactobacilli involved in the acidification of milk. Tuberculosis was then one of the most important infectious diseases, and his Hoagland Laboratory colleague Benjamin White had to take the cure at the Trudeau sanatorium; Avery therefore worked on tubercle bacilli. When he joined The Rockefeller Institute Hospital, lobar pneumonia was the problem under investigation in the Department of Respiratory Diseases; he therefore became a specialist in the bacteriology of pneumococci. The approach to the control of pneumonia at The Rockefeller Institute was through the development of therapeutic antisera; he therefore studied the immuno-chemical processes that might contribute to vaccination and serotherapy. The success of chemotherapy, first with the sulfa drugs in the late 1930s, then with penicillin in the 1940s, made the immunochemical approach less urgent, so that he could concentrate his efforts on the isolation of the substance responsible for the transformation of pneumococcal types; in this case, again, he did not create the transformation problem, but rather faced up to it because Griffith's discovery had threatened the doctrine of immunological specificity to which he had become commited.

The methods that he used in his research also were provided by his time and, in particular, by the scientific environment in which he worked. Both at the Hoagland Laboratory and at The Rockefeller Institute, most of his colleagues believed that biological phenomena are only complex expressions of physicochemical processes, and that physics and chemistry offer the only pathways leading to a real understanding of animate nature. In agreement with this view, Avery made it his scientific ideal to formulate pathological and biological problems in physicochemical terms, and to define chemically the substances and reactions involved in the phenomena that he studied.

His originality and creativity did not reside in the kinds of problems on which he worked or in the development of new laboratory methods, but in the intellectual style of his investigations. He accepted the practical problems that came his way, but he recognized and emphasized certain aspects of them that had large biological significance; he used conventional laboratory methods, but he designed original and artistic experiments. For example, the immunological specificity of bacterial strains was a widely recognized phenomenon when he began his immunochemical studies, but he gave it a broader and richer significance by relating specificity to certain anatomical structures of the microorganisms and to certain chemical configurations of these structures. Whereas the transformation of pneumococcal types was regarded by most microbiologists as an oddity of little interest, he had the persistence and the vision to convert type transformation into a precise and elegant laboratory model of a phenomenon with great significance for theoretical biology.

Persistance was one of Avery's most striking and useful assets, not only because it made him an effective investigator, but especially because he applied it unerringly to important problems. Pasteur was wont to tell his associates whenever an important phenomenon seemed to escape his control: "Let us do the same experiment over again; the essential is never to leave the subject."[6] This was Avery's attitude, as most strikingly demonstrated during the 10 years of heart-breaking failures that preceded the development of a reproducible method of type transformation in pneumococci.

For him, however, persistence implied more than the willingness to continue a line of experimentation against odds; it meant pursuing a problem beyond the point of initial success. This attitude conditioned his pragmatic philosophy of the experimental method, as he expressed it during a conversation he had around 1935 with the young Dr. Barry Wood, who had come to him for advice before beginning his research career. Scientific investigators, Avery told Dr. Wood, can be divided into two classes. There are those, the most numerous, "who go around picking up the surface nuggets, and wherever they can spot a surface nugget of gold they . . . grab it and put it in their collection." On the other hand, there is the more unusual investigator "who is not really interested in the surface nugget. He is much more interested in digging a deep hole in one place, hoping to hit a vein. And of course if he strikes a vein of gold he makes a tremendous advance."[7] This statement, made by Avery years before he had established the role of DNA as the bearer of hereditary characteristics, reveals how clearly he realized that persistence was probably one of his most important assets as an investigator.

His persistence also accounts for the fact that he remained scientifically productive into very late in life. According to the English mathematician G. H. Hardy, "A mathematician may still be competent enough at sixty, but it is useless to expect him to have original ideas." [8] Thomas Huxley is reputed to have gone even further, and stated that "a man of science past sixty does more harm than good." [9] William Osler once facetiously referred to the admirable scheme of a college into which, at sixty, men retired for a year of contemplation before a peaceful departure by chloroform.[10] These statements express the commonly held view that creativity in science decreases rapidly after early adulthood.

Avery was past 65 when he published the DNA work, which is commonly regarded as his greatest achievement. Therefore, his case seems to be an oddity in the annals of science. The fact is, however, that this achievement did not depend on "original ideas" as commonly thought of, and as understood, for example, in Hardy's phrase. The transformation of pneumococcal types had been known for 15 years; the isolation and identification of the substance responsible for transformation did not require originality of concepts, but rather the disciplined and critical application of known laboratory techniques. A similar situation is presented by the case of the English physicist Lord Rayleigh. He, also, remained productive in classical science until his late 60's, and gave an explanation that is applicable to Avery's continued scientific creativity. When asked to comment on Huxley's remark that a "man of science past sixty does more harm than good," Lord Rayleigh replied, "That may be, if he undertakes to criticize the work of younger men, but I do not see why it need be so if he sticks to the things he is conversant with." [11] This is, of course, exactly what Avery did. He had a deep knowledge of pneumococcal biology; he was familiar with the technical problems of type transformation; and he sensed that, in some way, these problems had broad theoretical significance. A large part of his creativity thus resided in his wisdom. He knew that by "digging a deep hole in one place" he had a good chance to hit a vein, even though he could not predict what he would discover.

Experimental Science as an Art Form

Avery's advice to Dr. Barry Wood was a picturesque way of acknowledging that persistence had been an essential factor in his own success, but there was much more than that to his genius as an investigator. Before deciding where to dig the hole, he spent much time surveying the terrain and cogitating about the worth of the enterprise. Furthermore, he tried hard to imagine beforehand what kind of vein would be worth exploiting.

While he was scrupulous to the extreme in the establishment of facts, he acted as if he did not believe that truth would automatically emerge from those facts. His approach to knowledge was not through compulsive scholarship and the accumulation of data, but rather through an imaginative vision of reality expressed in hypotheses derived from a few carefully selected facts.

All experimenters worth their salt go, of course, through the process of hypothesis-making in the course of their work; furthermore, all believe that a hypothesis can be useful, irrespective of its validity, because the very findings that show it to be erroneous commonly suggest new lines of investigation. However, this orthodox view of the experimental method, conceived as a continuous interplay and feedback between hypotheses and experimentation, does not do full justice to Avery's way of dealing with scientific problems. He was as much interested in constructing elegant mental models of the truth as in describing reality.

His formulation of scientific problems had some analogy to the wonderfully entertaining way he had of telling stories about matters of everyday life. These stories were very close to the truth, but differed from it in form, if not in spirit. They were made up of factual elements organized in such a way as to create a composition more interesting and more compelling than the actual occurrence. Similarly, he loved to create theoretical images out of the scientific facts provided by observation and experimentation. Throughout his scientific career, for example, he composed hypotheses in the form of short phrases, the meaning of which could almost be visualized from his choice of words. The following are a few of the word images on which he focused his thoughts, and around which he organized his experiments.

Antiblastic immunity: the metabolic processes of the host which inhibit the multiplication of parasites.

Host chemistry: the various chemical changes that occur in the body as a consequence of infection.

Specific soluble substances: the substances produced by the various types of pneumococci that determine the immunological specificity of each particular type.

Capsular antigen: the cellular complex of which each type of capsular substance is a part in virulent encapsulated pneumococci, and which is responsible for the ability to induce specific immunity.

Antigenic dissociation: the enzymatic processes caused either by pneumococci or by infected hosts that separate the capsular substance from the complex structure of which it is a part in the virulent encapsulated cell.

Rabbit virulence factor: a cellular substance, other than the capsular polysaccharide, which determines the ability of encapsulated pneumococci to cause disease in rabbits.

Transforming agent: the component of a pneumococcal cell that enables the cell to transfer its immunological specificity to other pneumococci.

As we have seen in preceding chapters, Avery converted many of these word images into laboratory operations that established the existence of each of the phenomena symbolized by the image. In several cases, his experimental studies led to the chemical isolation and identification of the substance responsible for the phenomenon.

The occasions when experimental findings revealed the factual basis of one of his word images and thus gave it a concrete meaning were, for Avery, moments of childlike pleasure that he wished to share with his colleagues and, indeed, with a broader public. On these occasions, The Professor became the showman.

The quality of his showmanship had much in common with the spectacular demonstrations staged by Pasteur during the early days of the controversies about the germ theory of fermentation and of disease. Best known is Pasteur's famous experiment at the farm at Pouilly le Fort, where he arranged for a widely publicized field demonstration of the fact that sheep can be protected against anthrax by vaccination. More similar to Avery's case, because on a smaller scale and for a more specialized audience, was the demonstration Pasteur staged before the Paris Academy of Medicine with four chickens of different plumage to convince his colleagues that these birds can be made susceptible to anthrax by lowering their body temperature.[12]

The spectacular demonstrative value of Pasteur's public performance depended, of course, upon his complete mastery of experimental conditions. Numerous prior trials had made it safe for him to eliminate all unnecessary details of the experiment and thus to increase its impact when it was performed in public. Avery, also, would first work out the precise requirements for a foolproof demonstration of the phenomenon he considered of importance, and then design tests as simple as compatible with providing irrevocable evidence. These final tests would include a few control animals or test tubes showing no effect whatever, a few others placed under such limiting conditions that the effect was apparent but minimal, and finally a few others in which the ideal conditions assured unequivocal and striking results, whether the phenomenon being demonstrated was acute death of an animal or its resistance to disease, formation of a precipitate in a test tube or its inhibition. As discussed in Chapter Five,

Avery called these simplified tests "protocol experiments," and he loved to perform them before colleagues and visitors.

These protocol experiments certainly had a meaning that transcended the pleasure he derived from the demonstration. They symbolized for him some of the values that had made him choose a life of science, in particular the serenity, security, and order that can be found in the world of experimentation, where much can be understood and controlled. Einstein has movingly expressed these values in the following words that are largely applicable to Avery:

> . . . one of the strongest motives that lead persons to art and science is flight from the everyday life, with its painful harshness and wretched dreariness, and from the fetters of one's own shifting desires. . . .
>
> With this negative motive there goes a positive one. Man seeks to form for himself, in whatever manner is suitable for him, a simplified and lucid image of the world, and so to overcome the world of experience by striving to replace it to some extent by this image. This is what the painter does, and the poet, the speculative philosopher, the natural scientist, each in his own way. Into this image and its formation, he places the center of gravity of his emotional life, in order to attain the peace and serenity that he cannot find within the narrow confines of swirling, personal experience.[13]

In his protocol experiments, Avery behaved much as artists do in their efforts to convey their response to the external world. Artists deal with limited aspects of reality, selecting from it only what they need to express an inner vision or concept. Furthermore, they deliberately impose on themselves limits as to their mode of expression—for example, a sonnet or a canvas of a particular shape and size. In the end, the value of the poem or the painting does not reside in the situation it describes, but in the poem itself or the painting itself—as a new creation and as a personal vision of reality. The frame placed around a picture symbolizes that the painter has elected to separate from the cosmos a fragment of nature and to make of it a self-sufficient entity through his own interpretation and vision. Similarly, the design of an experiment provides a pattern of reality controlled and shaped by the mind of the experimenter.

When Avery displayed a phenomenon with a few test tubes or animals, he gave an independent existence to a fragment of reality and created his view of scientific truth. For him, science was more than problem-solving or the accumulation of facts. It meant recognizing patterns in the apparent chaos of nature and composing the raw materials of nature into artistic creations.

ENVOI

The spirit of scientific research which emerged in American medicine around the turn of the century was incarnated in The Rockefeller Institute for Medical Research. A letter written by Simon Flexner shortly after retiring from his directorship of the Institute shows that Avery was for him a perfect expression of this spirit:

> . . . I regard it as one of the pieces of greatest good fortune for the Institute . . . that you came there so early in the Hospital's history and are still there to carry on your most important and original work, which no one else could possibly have done as you have done it. . . . There is no one that I have got more pleasure and stimulation in talking with than yourself. It was one of my privileges to have this understanding, intimate relation with you.[1]

Coming from so reserved a man as Simon Flexner, this letter is an extraordinary statement of what Avery meant to the Institute. Avery, on the other hand, knew that the Institute had been for him an ideal spiritual home, one in which he had discovered himself or, more exactly, made himself into what he wanted to become. In 1945, two years after his own official retirement, he wrote Flexner, "No words of mine can ever convey to you my gratitude for all you have done and *made possible* for me these many years"[2] (italics mine). What Flexner and the Institute had made possible was to cultivate in full freedom a few characteristics that gave a distinctive and unique quality to Avery's scientific style and personal life.

Avery was remarkable as a scientist by his ability to recognize important problems and by his mastery of the experimental method, but even more by his research style. Everything he did in his adult life had an artistic quality governed by a classical taste and a strict discipline. He shunned uncertainty, vagueness, and overstatement in scientific matters as much as in everyday life.

He did not have a robust enough temperament to deal effectively with complex, ill-defined situations, such as those commonly presented by clinical and social problems, but he had immense intellectual vigor in selecting from the confusion of natural occurrences the few facts most significant for the problems he elected to investigate, and he had the creative impulse to compose these facts into meaningful and elegant structures. His scientific compositions had, indeed, much in common with artistic creations, which do not imitate actuality, but transcend it and illuminate reality.

Avery applied disciplined creativeness both to his scientific work and to the development of his personality. He retained throughout his life the perceptive, intelligent, determined, and also impish and whimsical expression that had characterized him during his youth and college years. In adulthood and old age, however, his face radiated, in addition, tolerance, sympathy, wisdom, and a romantic inwardness. "At 50, everyone has the face he deserves." [3] This was especially true of Avery, whose adult face achieved a rich mellowness that testified to the prodigious control he exerted over all aspects of his temperament. He certainly believed with Montaigne that each of us can "discover in himself a pattern all his own" and that "to compose our character is our duty." [4] In the end, his most glorious masterpiece was the persona he created by cultivating at each phase of his intellectual and emotional development those aspects of his nature that made him function best in each particular situation.

Those who have known The Professor admire him for what he composed as a scientist; but they remember him even more vividly for the art with which he composed his character and his life.

REFERENCES

NOTE: *Throughout, BSD denotes annual report to the Board of Scientific Directors of The Rockefeller Institute.*

CHAPTER ONE

The Professor and the Institute

1. Blake, William. 1965. *The Marriage of Heaven and Hell.* Pinto, Vivian, Ed. New York: Schocken, p. 101.
2. Bacon, Francis. 1960. *The New Organon and Related Writings.* Anderson, F. H., Ed. New York: Liberal Arts Press, p. 96.
3. Pasteur, Louis. 1926. *Oeuvres de Pasteur.* Vol. 7. Paris: Masson et Cie., pp. 200, 427.
4. Corner, George. 1964. *A History of The Rockefeller Institute.* New York: The Rockefeller Institute Press, pp. 39–41.
5. *Ibid.,* p. 55.
6. *Ibid.,* p. 64.
7. *Ibid.,* p. 94.

CHAPTER TWO

From the Bedside to the Laboratory

1. Flexner, S. and Flexner, J. T. 1941. *William Henry Welch and the Heroic Age of American Medicine.* New York: Viking, p. 230.
2. Corner, G. W. 1965. *Two Centuries of Medicine: A History of the School of Medicine, University of Pennsylvania.* Philadelphia: J. B. Lippincott, Chapter 2. See also: Fleming, D. 1954. *William H. Welch and the Rise of Modern Medicine.* Boston: Little, Brown, p. 4.
3. Cushing, H. 1925. *The Life of Sir William Osler.* Vols. I and II. Oxford: Clarendon, p. 546.
4. Cohn, Alfred. 1948. *No Retreat from Reason.* New York: Harcourt, Brace, p. 34.
5. Flexner and Flexner. *Op. cit.,* p. 291.
6. *Ibid.*
7. *Ibid.,* p. 112.
8. Fleming, D. *Op. cit.,* p. 7.
9. Eggerth, Arnold. 1960. *The History of the Hoagland Laboratory.* New York: n.p.
10. *Ibid.,* p. 43.
11. *Ibid.,* p. 110.
12. Flexner and Flexner. *Op. cit.,* pp. 269–296. See also: Corner, G., 1964. *A History of the Rockefeller Institute.* New York: The Rockefeller Institute Press.

13. *Ibid.*, pp. 578–580.
14. Flexner and Flexner. *Op. cit.*, and Corner, *Op. cit.*
15. Corner, G. W. *Op. cit.*, facing title page.
16. Fleming, D. *Op. cit.*, p. 153.
17. Flexner and Flexner. *Op. cit.*, p. 289.
18. Bernard, Claude. 1885. *Leçons sur les phénomènes de la vie communs aux animaux et aux végétaux.* Paris: J. B. Bailliere.
19. Corner, G. W. *Op. cit.*, p. 89.
20. Flexner, Simon. 1939. *The Evolution and Organization of the University Clinic.* Oxford: Clarendon Press, p. 18.
21. *Ibid.*, p. 26.
22. Swift, Homer. 1928. The art and science of medicine. *Science* 68: 167. See also: Cohn, Alfred, 1931. *Medicine, Science and Art: Studies in Interrelations.* Chicago: Univ. of Chicago Press. Also: Rivers, T. 1950. Concepts and Methods of Medical Research. The George R. Siedenburg Memorial Lecture. In: *Frontiers in Medicine.* New York: Columbia Univ. Press, p. 120.
23. Benison, S. 1967. *Tom Rivers: Reflections on a Life in Medicine and Science.* Cambridge, Mass.: M.I.T. Press, p. 200.
24. *Ibid.*, p. 196.
25. Flexner, Abraham. 1910. *Medical Education in the United States and Canada.* Carnegie Foundation, Bulletin Number 4.
26. Flexner, Simon. *Op cit.*, p. 37.
27. de Kruif, Paul. *Op. cit.*, p. 16.
28. Flexner, Simon. *Op. cit.*, p. 37.
29. *Ibid.*, p. 35.

CHAPTER THREE

Chemistry in Medical Research

1. Corner, George. 1964. *A History of The Rockefeller Institute.* New York: The Rockefeller Institute Press, pp. 578–580.
2. Flexner, S., and Flexner, J. T. 1941. *William Henry Welch and the Heroic Age of American Medicine.* New York: Viking, p. 280.
3. *Ibid.*, p. 282.
4. *Ibid.*, p. 276.
5. *Ibid.*, p. 289.
6. *Ibid.*, p. 55.
7. *Ibid.*, p. 63.
8. Corner, G. W. *Op. cit.*, p. 13. See also: Hawthorne, Robert. 1974. Christian Archibald Herter, M.D. (1865–1910). *Perspect. Biol. Med.* 18: 24–39.
9. Flexner and Flexner. *Op. cit.*, p. 283.
10. *Ibid.*
11. *Ibid.*, p. 284.
12. Quoted in Dubos, R. 1945. *The Bacterial Cell.* Cambridge: Harvard Univ. Press, p. 229.

13. *Ibid.,* p. 92.
14. Fleming, Donald. 1964. Introduction to Loeb, Jacques, *The Mechanistic Conception of Life.* Cambridge: Harvard Univ. Press, p. viii.
15. *Ibid.* See also: Osterhout, W. J. V. 1928. Jacques Loeb. *J. Gen. Physiol.* 8: ix–lv. Also: Reingold, Nathan. 1962. Jacques Loeb, the Scientist. *Library Congr. Quart. J.* 19: 119–130.
16. Fleming, D. *Op. cit.,* p. xiii.
17. *Ibid.,* p. xviii.
18. Osterhout, W. J. V. *Op. cit.,* p. xviii.
19. *Ibid.,* p. li.
20. de Kruif, Paul. 1962. *The Sweeping Wind.* New York: Harcourt, Brace and World, p. 16.
21. Cohn, Alfred. 1948. *No Retreat from Reason.* New York: Harcourt, Brace, p. 263.
22. *Ibid.,* p. 264.
23. Osterhout, W. J. V. *Op. cit.,* p. xxiii.
24. Fleming, Donald. 1954. *William H. Welch and the Rise of Modern Medicine.* Boston: Little, Brown, p. 153.
25. Corner, G. W. *Op. cit.,* pp. 328–329.
26. Fleming, D. Introduction to Loeb, p. xli (see ref. 14, above).

CHAPTER FOUR

Avery's Personal Life

1. Avery, O. T. 1944. Karl Landsteiner. *J. Pathol. Bacteriol.* 56: 592.
2. Bernard, Claude. 1942. *Le Cahier Rouge.* Paris: Gallimard, p. 119.
3. Flexner, S., and Flexner, J. T. 1941. *William Henry Welch and the Heroic Age of American Medicine.* New York: Viking.
4. Unless otherwise noted, the material concerning the Avery family is in the personal collection of Mrs. Roy Avery, or The Manuscript Unit, Archives and Records Services Section, Tennessee State Library and Archives, Nashville, Tennessee, Accession Number 70-128.
5. *Buds and Blossoms,* Vol. X, Jan., 1886.
6. *Buds and Blossoms,* Vol. XI, Aug., 1887.
7. *Buds and Blossoms,* Vol. XV, May, 1891.
8. Rockefeller Family Archives, New York, Record Group I, March 12, 1891.
9. *Ibid.,* December 2, 1890.
10. *Ibid.,* December 29, 1893.
11. *Ibid.,* December 30, 1890.
12. Flynn, J. 1932. *God's Gold.* New York: Harcourt, Brace, p. 269. See also: Nevins, Allan. 1959. *John D. Rockefeller.* New York: Scribner's, p. 224.
13. *Buds and Blossoms,* Vol. XVI, Dec., 1892.
14. *Buds and Blossoms,* Vol. XVI, Apr., 1892.
15. Letter to author, dated October 6, 1975.
16. *Buds and Blossoms,* Vol. XVI, Apr., 1892.
17. *Ibid.*

18. *Buds and Blossoms,* Vol. XIII, June, 1889.
19. *Buds and Blossoms,* Vol. XVI, Apr., 1892.
20. Williams, Howard. 1969. *A History of Colgate University, 1819–1969.* New York: Van Nostrand Reinhold.
21. Fosdick, Harry Emerson. 1956. *The Living of These Days.* New York: Harper, p. 53.
22. *Ibid.,* p. 58.
23. *Salmagundi 1900.* Colgate University Yearbook, p. 28.
24. *Ibid.,* p. 31.
25. *Ibid.*
26. Colgate University Archives, Hamilton, New York.
27. *Colgate Alumni News,* August, 1965, p. 22.
28. Coburn, A. F. 1974. *Commitment Total.* New York: Walker, p. 186.
29. Dochez, A. R. 1958. *Oswald Theodore Avery. National Academy of Sciences Biographical Memoirs.* Vol. 32. New York: Columbia Univ. Press, p. 31.
30. Eggerth, Arnold. 1960. *The History of the Hoagland Laboratory.* New York: n.p., p. 125.
31. *Ibid.,* pp. 125–26.
32. Heidelberger, M., Kneeland, Y., Jr., and Price, K. 1971. *Alphonse Raymond Dochez. National Academy of Sciences Biographical Memoirs.* Vol. 42. New York: Columbia Univ. Press, p. 39.
33. MacLeod, Colin. September 29, 1965. Remarks delivered at the dedication of the Avery Gateway. Rockefeller University Archives.
34. Chesney, Alan. 1957. Oswald Theodore Avery. *J. Pathol. Bacteriol.,* 74: 454.
35. Letter to author dated November 7, 1975.
36. Letter to author, from Mrs. Margaret Brearley, dated November 14, 1975.
37. *Ibid.*
38. *Ibid.*

CHAPTER FIVE

Avery's Life in the Laboratory

1. Avery, O. T. 1949. Presentation of the Kober Medal Award to Dr. Alphonse R. Dochez. *Trans. Assoc. Am. Physicians* 62: 28.
2. Avery, O. T. 1946. Acceptance of the Kober Medal Award. *Trans. Assoc. Am. Physicians* 59: 45.
3. BSD. 1923–24. Vol. 12, p. 138.
4. MacLeod, Colin. 1957. Oswald Theodore Avery, 1877–1955. *J. Gen. Microbiol.* 17: 543.
5. McCarty, Maclyn. 1965. Oswald T. Avery and his scientific legacy. *Rockefeller University Review,* Sept.–Oct., p. 12.
6. Quoted in *Scientific Research,* October, 1967, n.p.
7. Loir, Adrien. 1938. *A l'ombre de Pasteur.* Paris: Le Mouvement Sanitaire, pp. 28, 50.

8. Hotchkiss, R. D. 1965. Oswald T. Avery. *Genetics* 51: 3.
9. *Ibid.*, p. 4.
10. *Ibid.*, p. 3.

CHAPTER SEVEN

The Lure of Antiblastic Immunity
and the Chemistry of the Host

1. Dochez, A. R., and Avery, O. T. 1916. Antiblastic immunity. *J. Exp. Med.* 23: 61–68.
2. BSD. April, 1915. Vol. 4, p. 108.
3. *Ibid.*, p. 73.
4. Dochez and Avery. *Op. cit.*, p. 67.
5. BSD. 1916. Vol. 4, p. 253.
6. Dochez and Avery. *Op. cit.*, p. 68.
7. BSD. 1916. Vol. 4, p. 252–253.
8. BSD. 1919. Vol. 7, p. 368.
9. Blake, F. G. 1917. Studies on antiblastic immunity. *J. Exp. Med.* 26: 563.
10. Barber, M. A. 1919. A study by the single cell method of the influence of homologous antipneumococcic serum on the growth rate of pneumococcus. *And* Antiblastic phenomena in active acquired immunity and in natural immunity to pneumococcus. *J. Exp. Med.* 30: 569–587, 589–596.
11. Marrack, J. R. 1950. Antibodies to Enzymes. *In:* Myrbäck, K., and Sumner, J. B., (Eds.). *The Enzymes.* Vol. I, New York: Academic Press. See also: Sevag, M. G. 1951. *Immuno-catalysis.* Springfield, Ill.: C. C Thomas.
12. Krebs, E. G., and Najjar, V. A. 1948. The inhibition of D-glyceraldehyde 3-phosphate dehydrogenase by specific antiserum. *J. Exp. Med.* 88: 569–577.
13. BSD. 1916. Vol. 4, p. 253.
14. BSD. April, 1920. Vol. 8, p. 101.
15. BSD. Oct., 1920. Vol. 8, p. 248.
16. BSD. 1923–24. Vol. 12, p. 130.
17. Hotchkiss, R. D. 1965. Oswald T. Avery. *Genetics* 51: 5.
18. Avery, O. T. 1941. The Commonwealth of Science. Presidential Address before the American Society of Bacteriologists. Unpublished typescript, personal property of author.
19. *Ibid.*, p. 1.
20. *Ibid.*, p. 8.
21. *Ibid.*, p. 8.

CHAPTER EIGHT

The Chemical Basis of Biological Specificity

1. Avery, O. T., Chickering, H. T., Cole, R., and Dochez, A. R. 1917. *Acute*

Lobar Pneumonia: Prevention and Serum Treatment. New York: The Rockefeller Institute for Medical Research, Monograph No. 7.

2. BSD. 1936–37. Vol. 25, pp. 314–15.
3. Dochez, A. R., and Avery, O. T. 1917. Soluble substance of pneumococcus origin in the blood and urine during lobar pneumonia. *Proc. Soc. Exp. Biol. Med.* 14: 126.
   Dochez, A. R., and Avery, O. T. 1917. The elaboration of specific soluble substance by pneumococcus during growth. *J. Exp. Med.* 26: 477.
4. BSD. April, 1917. Vol. 5, p. 137.
5. Dochez and Avery. *Op. cit., J. Exp. Med.,* p. 493.
6. BSD. 1922–23. Vol. 11, p. 145.
7. *Ibid.,* p. 146.
8. BSD. 1923–24. Vol. 12, p. 136.
9. *Ibid.,* p. 137.
10. *Ibid.,* p. 140.
11. *Ibid.,* p. 142.
12. BSD. 1924–25. Vol. 13, p. 306.
13. BSD. 1925–26. Vol. 14, p. 622.
14. *Ibid.,* p. 624.
15. BSD. 1924–25. Vol. 13, p. 312.
16. *Ibid.,* p. 310.
17. Goebel, Walther. 1975. The golden era of immunology at The Rockefeller Institute. *Perspect. Biol. Med.* 18: 419–426.
18. BSD. 1930–31. Vol. 19, pp. 412–413.
19. Landsteiner, Karl. 1962. *The Specificity of Serological Reactions.* New York: Dover, p. 176.
20. BSD. 1924–25. Vol. 13, p. 310.
21. Northrop, J. H., and Goebel, Walther, F. Crystalline pneumococcus antibody. *J. Gen. Physiol.* 32: 705.

CHAPTER NINE

The Complexities of Virulence

1. Avery, O. T. 1941. The Commonwealth of Science. Presidential Address before the American Society of Bacteriologists. Unpublished typescript, personal property of author.
2. BSD. 1930–31. Vol. 19, p. 406.
3. Avery, O. T., Chickering, H. T., Cole, R., and Dochez, A. R. 1917. *Acute Lobar Pneumonia: Prevention and Serum Treatment.* New York: The Rockefeller Institute for Medical Research, Monograph No. 7, p. 10.
4. *Ibid.,* p. 16.
5. BSD. 1923–24. Vol. 12, p. 339.
6. BSD. 1926–27. Vol. 15, p. 519.
7. *Ibid.*
8. MacLeod, C., and McCarty, M. 1942. The relation of a somatic factor to virulence of pneumococci. *J. Clin. Invest.* 21: 647.

CHAPTER TEN

Bacterial Variability

1. Pasteur, L. 1857. Mémoire sur la fermentation appelée lactique. *C. R. Acad. Sci.* 45: 913–16.

2. Cohn, F. 1866. *Ueber Bacterien, die kleinsten lebenden Wesen*. Berlin: Samml. gemeinverständl. wissenschaftl. Vorträge, hrsg. v. R. Virchow ü. Fr. v. Holtzendorff, no. 165.

3. Huxley, T. H. 1870. On the relations of Penicillium, Torula, and Bacterium. *Q. J. Microsc. Sci.* N. S. 10: 355–62.

4. Klebs, E. 1873. Beiträge zur Kenntniss der Micrococcen. *Arch. Exp. Pathol. Pharmakol.* 1: 31–64.

5. Lankester, E. Ray. 1873. On a peach-coloured bacterium – Bacterium rubescens. *Q. J. Microsc. Sci.* N. S. 13: 408–25.

6. Billroth, T. 1874. *Untersuchungen über die Vegetations formen von Coccobacteria septica und den Antheil, welchen sie an der Entstehung und Verbreitung der accidentellen Wundkrankheiten haben.* Berlin: Reimer.

7. Lister, J. 1873. A further contribution to the natural history of bacteria and the germ theory of fermentative changes. *Q. J. Microsc. Sci.* N. S. 13: 380–408.

———. 1876. A contribution to the germ theory of putrefaction and other fermentative changes and to the natural history of torulae and bacteria. *Trans. R. Soc. Edinb. (1872–6).* 27: 313–44.

8. Nägeli, C. von. 1877. *Die niederen Pilze in ihren Beziehungen zu den Infectionskrankheiten und der Gesundheitspflege.* Munich: R. Oldenbourg.

9. Cohn, F. 1876. Untersuchungen über Bacterien. IV. Beiträge zur Biologie der Bacillen. *Beitr. Biol. Pflanz.* 2 (No. 2): 248–76.

10. Koch, R. 1876. Die Aetiologie der Milzbrand-krankheit, begründet auf die Entwicklungsgeschichte des *Bacillus anthracis. Beitr. Biol. Pflanz.* 2: 277–310.

11. Pasteur, L. 1876. *Études sur la bière, ses maladies, causes qui les provoquent, procédé pour la rendre inaltérable, avec une théorie nouvelle de la fermentation.* Paris: Gauthier-Villars

12. Davaine, C. 1872. Recherches sur quelques questions relatives à la septicémie. *Bull. Acad. Méd.,* 2nd sér. 1: 907, 976.

13. Pasteur, L. 1880. De l'atténuation du virus du choléra des poules. *C. R. Acad. Sci.* 91: 673–80.

14. ———. 1881. De l'atténuation des virus et de leur retour à la virulence. *C. R. Acad. Sci.* 92: 429–35.

15. Firtsch, G. 1888. Untersuchungen über Variationserscheinungen bei *Vibrio proteus. Arch. Hyg.* 8: 369–401.

16. Beijerinck, M. 1901. Mutation bei Mikroben. *Versl. Afd. Natuurkunde, Akad. Wetensch (Amsterdam)* 9: 310.

17. Neisser, M. 1906. Ein Fall von Mutation nach De Vries bei Bakterien und andere Demonstrationen. *Cent. Bakt. I. Ref.* 38: 98.

18. Massini, R. 1907. Ueber einen in biologischer Bezeichnung interessanten Kolistamm *(Bacterium coli mutabile). Arch. Hyg.* 61: 250–292.

19. Arkwright, J. A. 1921. Variation in bacteria in relation to agglutination both by salts and by specific serum. *J. Pathol. Bacteriol.* 24: 36–60.
20. *Ibid.,* p. 55.
21. Dubos, R. 1945. *The Bacterial Cell.* Cambridge: Harvard Univ. Press.
22. ———. 1932. Factors affecting the yield of specific enzyme in cultures of the bacillus decomposing the capsular polysaccharide of type III pneumococcus. *J. Exp. Med.* 55: 377–391.
23. ———. 1940. The adaptive production of enzymes by bacteria. *Bacteriol. Rev.* 4: 1–16.
24. Spiegelman, S., and Campbell, A. 1956. The Significance of Induced Enzyme Formation. In: Green, D. E. (Ed.), *Currents in Biochemical Research 1956.* New York: Interscience, pp. 115–161.
25. Karstrom, H. 1937–38. Enzymatische Adaptation bei Mikroorganismen. *Ergeb. Enzymforsch.* 7: 350–376.
26. Jacob, F., and Monod, J. 1961. Genetic regulatory mechanisms in the synthesis of proteins. *J. Mol. Biol.* 3: 318–356.
27. Dubos, R. 1945. *The Bacterial Cell.* Cambridge: Harvard Univ. Press, pp. 137–143.
28. Ravin, A. 1961. The genetics of transformation. *Adv. Genet.* 10: 61–163.
29. Coburn, A. 1974. *Commitment Total.* New York: Walker, p. 167.
30. Elliott, S. (quoted in Olby, R. 1974. *The Path to the Double Helix.* Seattle: Univ. of Washington Press), p. 170.
31. H. D. W. 1941. Frederick Griffith. *Lancet* 1: 588.
32. *Ibid.,* p. 589.
33. Griffith, F. 1923. The influence of immune serum on the biological properties of pneumococci. In: *Reports on Public Health and Medical Subjects.* No. 18. *Bacteriological Studies.* London: H. M. S. O., pp. 1–13.
34. ———. 1928. The significance of pneumococcal types. *J. Hyg.* 27: 113–159.
35. H. D. W. *Op. cit.,* p. 588.
36. Griffith, F. 1922. Types of pneumococci obtained from cases of lobar pneumonia. *Reports on Public Health and Medical Subjects.* No. 13. *Bacteriological Studies.* London: H. M. S. O., p. 36.
37. ———. *Op. cit.,* p. 153.
38. *Ibid.,* p. 153.
39. Neufeld, F., and Levinthal, W. 1928. Beiträge zur Variabilität der Pneumokokken. *Z. Immunitätsforsch.* 55: 324–340.
40. Reimann, H. 1929. The reversion of R to S pneumococcus. *J. Exp. Med.* 49: 237–249.
41. BSD. 1926–27. Vol. 15, p. 269.
42. Dawson, M. 1930. The transformation of pneumococcal types. I and II. *J. Exp. Med.* 51: 99–122 and 123–147.
43. Dawson, M., and Sia, R. 1931. *In vitro* transformation of pneumococcal types. I and II. *J. Exp. Med.* 54: 681–699, 701–710.
44. Alloway, J. 1932. The transformation *in vitro* of R pneumococci into S forms of different specific types by the use of filtered pneumococcus extracts. *J. Exp. Med.* 55: 91–99.

———. 1933. Further observations on the use of pneumococcus extracts in effecting transformation of type *in vitro. J. Exp. Med.* 57: 265–278.

CHAPTER ELEVEN

Heredity and DNA

1. Hotchkiss, R. 1965. Oswald T. Avery, 1877–1955. *Genetics* 51: 5.
2. Dubos, R., and Thompson, R. 1938. The decomposition of yeast nucleic acid by a heat-resistant enzyme. *J. Biol. Chem.* 124: 501–510.
3. The following discussion is taken from BSD. 1940–41, Vol. 29, p. 141–146.
4. Avery, O. T., MacLeod, C., and McCarty, M. 1944. Studies on the chemical nature of the substance inducing transformation of pneumococcal types. *J. Exp. Med.* 79: 137–158.
5. *Ibid.,* p. 155.
6. Mirsky, A. 1947. Contribution to the discussion of Boivin's paper. *Cold Spring Harbor Symp. Quant. Biol.* 12: 15–16.
7. Darlington, quoted in Olby, *op. cit.,* p. 192.
8. BSD. 1946–47. Vol. 35, p. 127.
9. McCarty, M. 1945. Reversible inactivation of the substance inducing transformation of pneumococcal types. *J. Exp. Med.* 81: 501–514.
10. ———. 1946. Purification and properties of desoxyribonuclease isolated from beef pancreas. *J. Gen. Physiol.* 29: 123–139.
11. ———. 1946. Chemical nature and biological specificity of the substance inducing transformation of pneumococcal types. *Bacteriol. Rev.* 10: 63–71.
12. McCarty, M., and Avery, O. 1946. Studies on the chemical nature of the substance inducing transformation of pneumococcal types. II. Effect of desoxyribonuclease on the biological activity of the transforming substance. *J. Exp. Med.* 83: 89–96.
13. ———. 1946. Studies on the chemical nature of the substance inducing transformation of pneumococcal types. III. An improved method for the isolation of the transforming substance and its application to pneumococcus types II, III, and VI. *J. Exp. Med.* 83: 97–104.
14. McCarty, M., Taylor, H. E., and Avery, O. T. 1946. Biochemical studies of environmental factors essential in transformation of pneumococcal types. *Cold Spring Harbor Symp. Quant. Biol.* 11: 177–183.
15. MacLeod, C. M., and Krauss, M. R. 1947. Stepwise intra-type transformation of pneumococcus from R to S by way of a variant intermediate in capsular polysaccharide production. *J. Exp. Med.* 86: 439–453.
16. Hotchkiss, R. D. 1948. Etudes chimiques sur le facteur transformant du pneumocoque. In: Lwoff, A. (Ed.). *Les unités biologiques douées de continuité génétique.* Paris: C. N. R. S., pp. 57–65.
17. Austrian, R., and MacLeod, C. M. 1949. Acquisition of M protein by pneumococci through transformation reactions. *J. Exp. Med.* 89: 451–460.
18. Taylor, H. E. 1949. Transformations réciproques des formes R et ER chez le pneumocoque. *C. R. Acad. Sci.* 228: 1258–1259.

19. Taylor, H. E. 1949. Additive effects of certain transforming agents from some variants of pneumococcus. *J. Exp. Med.* 89: 399–424.
20. Ephrussi-Taylor, H. E. 1951. Transformations allogènes du pneumocoque. *Exp. Cell Res.* 2: 589–607.
21. ———. 1951. Genetic aspects of transformations of pneumococci. *Cold Spring Harbor Symp. Quant. Biol.* 16: 445–456.
22. Hotchkiss, R. D. 1951. Transfer of penicillin resistance in pneumococci by the desoxyribonucleate derived from resistant cultures. *Cold Spring Harbor Symp. Quant. Biol.* 16: 457–461.
23. Hotchkiss, R. D., and Ephrussi-Taylor, H. E. 1951. Use of serum albumin as source of serum factor in pneumococcal transformation. *Fed. Proc.* 10: 200.
24. Hotchkiss, R. D. 1952. The role of desoxyribonucleates in bacterial transformations. In: McElroy, W., and Glass, B. (Eds.). *Phosphorus Metabolism.* Vol. II. Baltimore: Johns Hopkins Univ. Press, pp. 426–436.
25. Ephrussi-Taylor, H. E. 1954. A new transforming agent determining pattern of metabolism and glucose and lactic acids in pneumococcus. *Exp. Cell Res.* 6: 94–116.
26. Hotchkiss, R. D. 1954. Cyclical behavior in pneumococcal growth and transformability occasioned by environmental changes. *Proc. Natl. Acad. Sci.* 40: 49–55.
27. Hotchkiss, R. D., and Marmur, J. 1954. Double marker transformations as evidence of linked factors in desoxyribonucleate transforming agents. *Proc. Natl. Acad. Sci.* 40: 55–60.
28. Ephrussi-Taylor, H. E. 1955. Current status of bacterial transformations. *Adv. Virus Res.* 3: 275–307.
29. Hotchkiss, R. D. 1955. The Biological Role of the Deoxypentose Nucleic Acids. In: Chargaff, E., and Davidson, J. N. (Eds.), *Nucleic Acids.* Vol. II. New York: Academic Press, pp. 435–473.
30. ———. 1955. Bacterial transformation. *J. Cell. Comp. Physiol.* 45: (Suppl. 2) 1–14.
31. Marmur, J., and Hotchkiss, R. D. 1955. Mannitol metabolism a transferable property of pneumococcus. *J. Biol. Chem.* 214: 382–396.
32. Hotchkiss, R. D. 1956. The Genetic Organization of the Deoxyribonucleate Units Functioning in Bacterial Transformations. In: Gaebler, O. H. (Ed.). *Enzymes: Units of Biological Structure and Function.* New York: Academic Press, pp. 119–130.
33. MacLeod, C. M., and Krauss, M. R. 1956. Transformation reactions with two non-allelic R mutants of the same strain of pneumococcus type VIII. *J. Exp. Med.* 103: 623–638.
34. Ephrussi-Taylor, H. E. 1957. X-ray inactivation studies on solutions of transforming DNA from pneumococcus. In: McElroy, W., and Glass, B. (Eds.), *Chemical Basis of Heredity.* Baltimore: Johns Hopkins Univ. Press, pp. 299–320.
35. Fox, M. S., and Hotchkiss, R. D. 1957. Initiation of bacterial transformation. *Nature* 179: 1322–1325.
36. Hotchkiss, R. D. 1957. Criteria for Quantitative Genetic Transformations of Bacteria. In: *Chemical Basis of Heredity, op. cit.,* pp. 321–325.

37. Ephrussi-Taylor, H. E. 1958. The Mechanism of Desoxyribonucleic Acid-induced Transformations. In: Tunevalle, G. (Ed.). *Recent Progress in Microbiology*. Springfield, Illinois: C. C Thomas, pp. 51-68.

38. Hotchkiss, R. D. 1958. Size limitations governing the incorporation of genetic material in bacterial transformations and other nonreciprocal recombinations. *Symp. Soc. Exp. Biol.* 12: 49-59.

39. Hotchkiss, R. D., and Evans, A. H. 1958. Analysis of the complex sulfonamide resistance locus of pneumococcus. *Cold Spring Harbor Symp. Quant. Biol.* 23: 85-97.

40. Ephrussi-Taylor, H. E. 1960. On the biological functions of deoxyribonucleic acid. *Symp. Soc. Gen. Microbiol.* 10: 132-154.

41. ———. 1960. L'état du DNA transformant au cours des premières phases de la transformation bactérienne. *C. R. Soc. Biol.* 154: 1951-1955.

42. ———. 1961. Recombination Analysis in Microbial Systems. In: *Growth in Living Systems*. New York: Basic Books.

43. Lwoff, A. 1949. Unités biologiques douées de continuité génétique. *Colloq. Int. Cent. Natl. Rech. Sci.* No. 8, p. 202.

44. Hotchkiss, R. D. 1966. Gene, Transforming Principle, and DNA. In: Cairns, J., Stent, G., and Watson, J. (Eds.). *Phage and the Origins of Molecular Biology*. Cold Spring, N. Y.: Cold Spring Harbor Laboratory of Quantitative Biology, p. 194.

45. BSD. 1946-47. Vol. 35, pp. 126-127.

46. Hotchkiss, R. D. 1966. *Op. cit.*, pp. 193-194.

47. ———. 1965. Oswald T. Avery, 1877-1955. *Genetics* 51: 1 10, p. 6.

48. Fleming, D. 1969. Emigré Physicists and the Biological Revolution. In: Fleming, D., and Bailyn, B. (Eds.). *The Intellectual Migration*. Cambridge: Harvard Univ. Press, pp. 152-189.

49. Olby, R. 1974. Intellectual Migrations. In: *The Path to the Double Helix*. Seattle: Univ. of Washington Press, pp. 223-320.

50. Stent, G. S. 1972. Prematurity and uniqueness in scientific discovery. *Sci. Am.* 227: 84-93.

51. Dunn, L. C. 1951. *Genetics in the Twentieth Century: Essays on the Progress of Genetics during its First Fifty Years*. New York: Macmillan.

52. Stent, G. S. 1972. *Op. cit.*, p. 84.

53. Dobzhansky, T., quoted in Olby, 1974, *op. cit.*, p. 189.

54. Dobzhansky, T. 1941. *Genetics and the Origin of Species*. New York: Columbia University Press.

55. Hutchinson, G. E. 1945. The biochemical genetics of pneumococcus. *Am. Sci.* 33: 56-57.

56. Marshak, A., and Walker, A. C. 1945. Mitosis in regenerating liver. *Science* 101: 94-95.

57. Wright, S. 1945. Physiological aspects of genetics. *Ann. Rev. Physiol.* 1: 79, 83.

58. Beadle, G. W. 1948. Genes and biological enigmas. *Am. Sci.* 36: 71.

59. Burnet, Sir F. M. 1968. *Changing Patterns: An Atypical Biography*. London: Heinemann, p. 81.

60. Lwoff, A. 1949. *Op. cit.*

61. Dale, Sir H. 1946. Address of the President. *Proc. Royal Soc. (London).* 185A: 128.
62. Mirsky, A. 1947. Contribution to the discussion of Boivin's paper. *Cold Spring Harbor Symp. Quant. Biol.* 12: 15–16.
63. Chargaff, A. Quoted in Olby. *Op. cit.,* p. 211.
64. Watson, J. D. 1968. *The Double Helix.* New York: Atheneum, pp. 23, 48.
65. Stent, G. S. 1972. *Op. cit.,* p. 84.
66. Benzer, S. 1966. Adventures in the rII Region. In: Cairns, J., et al. (Eds.). *Phage and the Origins of Molecular Biology. Op. cit.,* p. 158.
67. Kalckar, H. M. 1966. High Energy Bonds: Optional or Obligatory. In: *Phage and the Origins of Molecular Biology. Op. cit.,* p. 46.
68. Stent, G. 1966. Waiting for the Paradox (Introduction). In: *Phage and the Origins of Molecular Biology. Op. cit.,* p. 4.
69. Hotchkiss, R. D. 1965. *Op. cit.,* p. 2.
70. Watson, J. D., and Crick, F. 1953. Molecular structure of nucleic acids. A structure for deoxyribose nucleic acid. *Nature* 171: 737–738.
    ———. 1953. Genetical implications of the structure of deoxyribonucleic acid. *Nature* 171: 964–967.
71. Schuck, H., et al. 1962. *Nobel, the Man and his Prizes.* New York: Elsevier, p. 281.

CHAPTER TWELVE

As I Remember Him

1. Flexner, S., and Flexner, J. T. 1941. *William Henry Welch and the Heroic Age of American Medicine.* New York: Viking, pp. 455–56.
2. Dr. Stuart Elliott, private communication.
3. Avery, O. T. 1941. The Commonwealth of Science. Presidential Address before the American Society of Bacteriologists. Unpublished typescript, personal property of author. All following quotations are taken from this typescript.
4. Magee, Bryan. 1973. *Popper.* London: Collins, Chapter 3.
5. Medawar, P. B. 1967. *The Art of the Soluble.* London: Methuen.
6. Pasteur, Louis. 1926. *Oeuvres de Pasteur.* Vol. 7. Paris: Masson et Cie., pp. 363–64, 390.
7. Wood, W. Barry, Jr. 1971. "Leaders in American Medicine," audiovisual memoir T/V2107. Alpha Omega Alpha Honor Medical Society and National Library of Medicine/National Medical Audiovisual Center.
8. Quoted in Chandrasekhar, S. 1975. Shakespeare, Newton and Beethoven, or Patterns of Creativity. The Nora and Edward Ryerson Lecture. Chicago: University of Chicago Center for Policy Study, p. 28.
9. *Ibid.,* p. 29.
10. Cushing, H. 1925. *The Life of Sir William Osler.* Oxford: Clarendon. Vol. I., p. 669.
11. Chandrasekhar. *Op. cit.,* p. 33.

12. Dubos, R. 1976. *Louis Pasteur, Free Lance of Science.* New York: Scribner's (reprint).
13. Einstein, A. 1954. *Ideas and Opinions of Albert Einstein.* New York: Crown, pp. 224–227.

ENVOI

1. Simon Flexner Papers. American Philosophical Society Library, Philadelphia, letter dated January 20, 1936.
2. *Ibid.,* letter dated April 9, 1945.
3. Orwell, George. 1968. *The Collected Essays, Journalism and Letters.* IV. *In Front of Your Nose 1945–1950.* Orwell, Sonia, and Angus, Ian (Eds.). New York: Harcourt, Brace and World, p. 515. See also: Camus, Albert. 1962. *La Chute.* Paris: Pleiade, p. 1502.
4. Montaigne, Michel de. 1958. *The Complete Essays* (Translated by D. Frame). Palo Alto: Stanford Univ. Press, p. 615.

APPENDIX VI

1. Ravin, A. 1961. The genetics of transformation. *Adv. Genet.* 10: 61–163.
2. *Ibid.*
3. Chargaff, E. 1950. Chemical specificity of nucleic acids and the mechanism of their enzymatic degradation. *Experientia* 6: 201–209.
   ———. 1957. The Base Composition of Desoxyribonucleic Acid and Pentose Nucleic Acid in Various Species. In: McElroy, W., and Glass, B. (Eds.). *A Symposium on the Chemical Basis of Heredity.* Baltimore: Johns Hopkins Press, pp. 521–527.
4. Chargaff, quoted in Robert Olby. 1974. *The Path to the Double Helix.* Seattle: Univ. of Washington Press, p. 211.
5. Robinow, C. 1942. A study of the nuclear apparatus of bacteria. *Proc. R. Soc. Lond. B. Biol. Sci.* 130: 299–324.
   ———. 1944. Cytological observations on *Bact. coli, Proteus vulgaris* and various aerobic spore-forming bacteria with special reference to the nuclear structures. *J. Hyg.* 43: 413–423.
6. Tulasne, R. 1947. Sur la mise en évidence du noyau des cellules bactériennes. *C. R. Séances Soc. Biol.* 141: 411–413.
7. Tulasne, R., and Vendrely, R. 1947. Demonstration of bacterial nuclei with ribonuclease. *Nature* 160: 225–226.
8. Hershey, A., and Chase, M. 1952. Independent functions of viral proteins and nucleic acid in growth of bacteriophage. *J. Gen. Physiol.* 36: 39–56.
9. Boivin, A., Vendrely, R., and Vendrely, C. 1948. L'acide desoxyribo-nucléique du noyau cellulaire, dépositaire des caractères héréditaires; arguments d'ordre analytique. *C. R. Hebd. Séances Acad. Sci., Paris* 226: 1061–1063.

10. Mirsky, A., and Ris, H. 1949. Variable and constant components of chromo-somes. *Nature* 163: 666–667.
11. Luria, S., and Delbrück, M. 1943. Mutation of bacteria from virus sensitivity to virus resistance. *Genetics* 28: 491–511.
12. Tatum, E., and Lederberg, J. 1947. Gene recombination in the bacterium *Escherichia coli. J. Bacteriol.* 53: 673–684.
13. Zinder, N., and Lederberg, J. 1952. Genetic exchange in Salmonella. *J. Bacteriol.* 64: 679–699.
14. Zinder, N. 1955. Bacterial transduction. *J. Cell. Comp. Physiol.* 45 (Suppl. 2): 23.
15. Lederberg, J. 1956. Genetic transduction. *Am. Sci.* 44: 264–280.
16. Ravin, A. 1960. The origin of bacterial species. *Bacteriol. Rev.* 24: 201–220.
———. 1961. The genetics of transformation. *Adv. Genet.* 10: 61–163.

CHRONOLOGIES

I · SOME EVENTS OF AVERY'S LIFE
ARRANGED IN CHRONOLOGICAL ORDER

Oct. 21, 1877	Born in Halifax
1887	His family moves to New York City
1893	Graduates from New York City Male Grammar School
1893–1896	Attends Colgate Academy in Hamilton, New York
1896–1900	B.A. from Colgate University with emphasis on humanistic studies and public speaking
1900–1904	M.D. from Columbia University College of Physicians and Surgeons in New York City
1904–1907	Medical practice (general surgery) in New York City
c. 1906	Given a research grant from New York City Board of Health to work on opsonic index
	Worked on milk pasteurization in bacteriological laboratory at the Sheffield Dairy Company, Brooklyn
1907–1913	Associate Director of bacteriological department at the Hoagland Laboratory, Brooklyn
1913–1948	The Rockefeller Institute for Medical Research, New York
	1913–1915 Assistant, Department of Hospital
	1915–1919 Associate, Department of Hospital
	1919–1923 Associate Member
	1923–1943 Member
	1943–1948 Emeritus Member
1917	Private, U.S. Army
1918	Acquires American citizenship
	Captain, U.S. Army
1948	Leaves New York for final retirement in Nashville, Tennessee
Feb. 20, 1955	Dies of cancer of the liver
	Buried in Mt. Olivet Cemetery, Nashville

Honorary Degrees

1921	Sc.D., Colgate University
1933	L.L.D., McGill University
1947	Sc.D., New York University
1950	Sc.D., University of Chicago
1954	Sc.D., Rutgers University

Awards

1930   Joseph Mather Smith Prize, Columbia University
1932   John Phillips Memorial Medal, American College of Physicians
1933   Paul Ehrlich Gold Medal (Germany)
1944   Medal of the New York Academy of Medicine
1945   Copley Medal, Royal Society of London
1946   Kober Medal, Association of American Physicians
1946   Charles Mickle Fellowship, University of Toronto
1947   Lasker Award, American Public Health Association
1949   Passano Award, Passano Foundation
1950   Pasteur Gold Medal, Swedish Medical Society of Stockholm

Scientific Organizations

Domestic
American Academy of Arts & Sciences
American Public Health Association
American Association for the Advancement of Science
American Association of Immunologists (President, 1923)
American Association of Pathologists and Bacteriologists
    (President, 1934)
American Society of Clinical Investigation
Association of American Physicans
Harvey Society
National Academy of Sciences
New York Academy of Medicine
Society for Experimental Biology and Medicine
Society of American Bacteriologists (President, 1942)
Society of Experimental Pathology

Foreign
Académie Royale de Médecine de Belgique
Der Norski Videnskaps Academi, Oslo
Pathological Society of Great Britian and Ireland
Royal Danish Academy of Science and Letters
Royal Society of London
Société Philomatique de Paris
Society of General Microbiology, England

Other
1942   Consultant to Secretary of War and Member of Board for Study
    and Control of Epidemic Diseases, U.S. Army.
1943   Member of Sub-Committee on Infectious Diseases, National
    Research Council
1948   Consultant and Member of Commission on Streptococcal Dis-
    eases, Epidemiological Board of the Armed Forces

II · SCIENTIFIC PUBLICATIONS OF AVERY
AND HIS COLLABORATORS

Work done at The Hoagland Laboratory

1909

White, B., and Avery, O. T. The treponema pallidum; observations on its occurrence and demonstration in syphilitic lesions. *Arch. Int. Med.* 3:411.

1910

Potter, N. B., and Avery, O. T. Opsonins and vaccine therapy. In: *Modern Treatment,* edited by Hare. Philadelphia and New York, Vol. I, p. 515.

White, B., and Avery, O. T. Observations on certain lactic acid bacteria of the so-called Bulgaricus type. *Cbl. Bakt.,* Abt. II. 25:161.

Ager, L. C., and Avery, O. T. A case of influenza meningitis. *Arch. Pediat.* 24:284.

White, B., and Avery, O. T. Concerning the bacteriemic theory of tuberculosis. *J. Med. Res.* 23:95.

1912

White, B., and Avery, O. T. The action of certain products obtained from the tubercle bacillus. A. Cleavage products of tuberculo-protein obtained by the method of Vaughan. Communication I. The poisonous substance. *J. Med. Res.* 26:317.

1913

Avery, O. T., and Lyall, H. W. Concerning secondary infection in pulmonary tuberculosis. *J. Med. Res.* 28:111.

White, B., and Avery, O. T. Some immunity reactions of edestin. The biological reactions of the vegetable proteins. III. *J. Inf. Dis.* 13:103.

1914

North, C. E., White, B., and Avery, O. T. A septic sore throat epidemic in Cortland and Homer, N. Y. *J. Inf. Dis.* 14:124.

Work done at The Rockefeller Institute for Medical Research

1915

Dr. Avery became a Member of the Institute in 1923. From then on, the names entered in parentheses for each academic year (July 1 to June 30) are those listed in the Annual Report to the Board of Scientific Directors. The list includes the departmental members of the scientific staff (M.D.'s, Ph.D.'s, and Guest Investigators) during the designated years; it does not include laboratory technicians or other helpers. (Inconsistencies in the original lists are reproduced here.)

Dochez, A. R., and Avery, O. T. Varieties of pneumococcus and their relation to lobar pneumonia. *J. Exp. Med.* 21:114.

Avery, O. T. The distribution of the immune bodies occurring in antipneumococcus scrum. *J. Exp. Med.* 21:133.

Dochez, A. R., and Avery, O. T. The occurrence of carriers of disease-producing types of pneumococcus. *J. Exp. Med.* 22:105.

Avery, O. T. A further study on the biologic classification of pneunococci. *J. Exp. Med.* 22:804.

1916

Dochez, A. R., and Avery, O. T. Antiblastic immunity. *J. Exp. Med.* 23:61.
Dochez, A. R., and Avery, O. T. Soluble substance of pneumococcus origin in the blood and urine during lobar pneumonia. *Proc. Soc. Exp. Biol. and Med.* 14:126.

1917

Avery, O. T., Chickering, H. T., Cole, R., and Dochez, A. R. Acute lobar pneumonia; prevention and serum treatment. *Monographs of The Rockefeller Institute for Medical Research,* No. 7, N. Y.
Dochez, A. R., and Avery, O. T. The elaboration of specific soluble substance by pneumococcus during growth. *J. Exp. Med.* 26:477. *Trans. Assoc. Amer. Phys.* 32:281.

1918

Avery, O. T. Determination of types of pneumococcus in lobar pneumonia: a rapid cultural method. *J. Amer. Med. Assoc.* 70:17.
Dernby, K. G., and Avery, O. T. The optimum hydrogen ion concentration for the growth of pneumococcus. *J. Exp. Med.* 28:345.
Avery, O. T. A selective medium for *B. influenzae.* Oleate-hemoglobin agar. *J. Amer. Med. Assoc.* 71:2050.

1919

Avery, O. T., and Cullen, G. E. The use of the final hydrogen ion concentration in differentiation of streptococcus haemolyticus of human and bovine types. *J. Exp. Med.* 29:215.
Dochez, A. R., Avery, O. T., and Lancefield, Rebecca C. Studies on the biology of streptococcus. I. Antigenic relationships between strains of streptococcus haemolyticus. *J. Exp. Med.* 30:179.
Avery, O. T., and Cullen, G. E. Hydrogen ion concentration of cultures of pneumococci of the different types in carbohydrate media. *J. Exp. Med.* 30:359.
Avery, O. T., Dochez, A. R., and Lancefield, Rebecca C. Bacteriology of streptococcus hemolyticus. *Ann. Otol. Rhinol. Laryngol.* 28:350.

1920

Avery, O. T., and Cullen, G. E. Studies on the enzymes of pneumococcus. I. Proteolytic enzymes. *J. Exp. Med.* 32:547.
Avery, O. T., and Cullen, G. E. Studies on the enzymes of pneumococcus. II. Lipolytic enzymes: esterase. *J. Exp. Med.* 32:571.
Avery, O. T., and Cullen, G. E. Studies on the enzymes of pneumococcus. III. Carbohydrate-splitting enzymes: invertase, amylase, and inulase. *J. Exp. Med.* 32:583.

1921

Thjötta, T., and Avery, O. T. Growth accessory substances in the nutrition of bacteria. *Proc. Soc. Exp. Biol. and Med.* 18:197.
Thjötta, T., and Avery, O. T. Studies on bacterial nutrition. II. Growth accessory substances in the cultivation of hemophilic bacilli. *J. Exp. Med.* 34:97.
Thjötta, T., and Avery, O. T. Studies on bacterial nutrition. III. Plant tissue, as a

source of growth accessory substances, in the cultivation of *Bacillus influenzae. J. Exp. Med.* 34:455.

Avery, O. T., and Morgan, H. J. The effect of the accessory substances of plant tissue upon growth of bacteria. *Proc. Soc. Exp. Biol. and Med.* 19:113.

1923–24

(with M. Heidelberger, H. J. Morgan, J. M. Neill)

Heidelberger, M., and Avery, O. T. The specific soluble substance of pneumococcus. *Proc. Soc. Exp. Biol. and Med.* 20:434.

Avery, O. T., and Heidelberger, M. Immunological relationships of cell constituents of pneumococcus. *Proc. Soc. Exp. Biol. and Med.* 20:435.

Heidelberger, M., and Avery, O. T. The soluble specific substance of pneumococcus. *J. Exp. Med.* 38:73.

Avery, O. T., and Heidelberger, M. Immunological relationships of cell constituents of pneumococcus. *J. Exp. Med.* 38:81.

Avery, O.T., and Cullen, G. E. Studies on the enzymes of pneumococcus. IV. Bacteriolytic enzyme. *J. Exp. Med.* 38:199.

Avery, O. T., and Morgan, H. J. Studies on bacterial nutrition. IV. Effect of plant tissue upon growth of pneumococcus and streptococcus. *J. Exp. Med.* 38:207.

Avery, O. T., and Morgan, H. J. Influence of an artificial peroxidase upon the growth of anaerobic bacilli. *Proc. Soc. Exp. Biol. and Med.* 21:59.

1924–1925

(with M. Heidelberger and W. F. Goebel, W. S. Tillett, L. A. Julianelle)

Avery, O. T., and Morgan, H. J. The occurrence of peroxide in cultures of pneumococcus. *J. Exp. Med.* 39:275.

Avery, O. T., and Morgan, H. J. Studies on bacterial nutrition. V. The effect of plant tissue upon the growth of anaerobic bacilli. *J. Exp. Med.* 39:289.

Morgan, H. J., and Avery, O. T. Growth-inhibitory substances in pneumococcus cultures. *J. Exp. Med.* 39:335.

Avery, O. T., and Neill, J. M. Studies on oxidation and reduction by pneumococcus. I. Production of peroxide by anaerobic cultures of pneumococcus on exposure to air under conditions not permitting active growth. *J. Exp. Med.* 39:347.

Avery, O. T., and Neill, J. M. Studies on oxidation and reduction by pneumococcus. II. The production of peroxide by sterile extracts of pneumococcus. *J. Exp. Med.* 39:357.

Avery, O. T., and Neill, J. M. Studies on oxidation and reduction by pneumococcus. III. Reduction of methylene blue by sterile extracts of pneumococcus. *J. Exp. Med.* 39:543.

Avery, O. T., and Neill, J. M. Studies on oxidation and reduction by pneumococcus. IV. Oxidation of hemotoxin in sterile extracts of pneumococcus. *J. Exp. Med.* 39:745.

Neill, J. M., and Avery, O. T. Studies on oxidation and reduction by pneumococcus. V. The destruction of oxyhemoglobin by sterile extracts of pneumococcus. *J. Exp. Med.* 39:757.

Heidelberger, M., and Avery, O. T. The soluble specific substance of pneumococcus. Second paper. *J. Exp. Med.* 40:301.

Neill, J. M., and Avery, O. T. Studies on oxidation and reduction by pneumococcus. VI. The oxidation of enzymes in sterile extracts of pneumococcus. *J. Exp. Med.* 40:405.

Neill, J. M., and Avery, O. T. Studies on oxidation and reduction by pneumococcus. VII. Enzyme activity of sterile filtrates of aerobic and anaerobic cultures of pneumococcus. *J. Exp. Med.* 40:423.

1925–1926

(with M. Heidelberger and W. F. Goebel)

Neill, J. M., and Avery, O. T. Studies on oxidation and reduction by pneumococcus. VIII. Nature of oxidation-reduction systems in sterile pneumococcus extracts. *J. Exp. Med.* 40:285.

Avery, O. T., and Morgan, H. J. Immunological reactions of isolated carbohydrate and protein of pneumococcus. *J. Exp. Med.* 42:347.

Avery, O. T., and Neill, J. M. The antigenic properties of solutions of pneumococcus. *J. Exp. Med.* 42:355.

Avery, O. T., and Heidelberger, M. Immunological relationships of cell constituents of pneumococcus. Second paper. *J. Exp. Med.* 42:367.

Heidelberger, M., Goebel, W. F., and Avery, O. T. The soluble specific substance of a strain of Friedländer's bacillus. Paper I. *J. Exp. Med.* 42:701.

Avery, O. T., Heidelberger, M., and Goebel, W. F. The soluble specific substance of Friedländer's bacillus. Paper II. Chemical and immunological relationships of pneumococcus Type II and of a strain of Friedländer's bacillus. *J. Exp. Med.* 42:709.

Heidelberger, M., Goebel, W. F., and Avery, O. T. The soluble specific substance of pneumococcus. Third paper. *J. Exp. Med.* 42:727.

Heidelberger, M., Goebel, W. F., and Avery, O. T. The soluble specific substance of a strain of Friedländer bacillus. *Proc. Soc. Exp. Biol. and Med.* 23:1.

Avery, O. T., Heidelberger, M., and Goebel, W. F. Immunological behaviour of the "E" strain of Friedländer bacillus and its soluble specific substance. *Proc. Soc. Exp. Biol. and Med.* 23:2.

Neill, J. M. Studies on the oxidation-reduction of hemoglobin and methemoglobin. I. The changes induced by pneumococci and by sterile animal tissue. *J. Exp. Med.* 41:299.

Neill, J. M. Studies on the oxidation-reduction of hemoglobin and methemoglobin. II. The oxidation of hemoglobin and reduction of methemoglobin by anaerobic bacilli and by sterile plant tissue. *J. Exp. Med.* 41:535.

Neill, J. M. Studies on the oxidation-reduction of hemoglobin and methemoglobin. III. The formation of methemoglobin during the oxidation of autoxidizable substances. *J. Exp. Med.* 41:551.

Neill, J. M. Studies on the oxidation-reduction of hemoglobin and methemoglobin. IV. The inhibition of "spontaneous" methemoglobin formation. *J. Exp. Med.* 41:561.

1926–1927

In the Annual Report for 1926–27, no publications are listed for any department. The publications listed on this page are taken from "The Semi-Annual List of the Publications of the Staff of The Rockefeller Institute for Medical Research," May 1926–November, 1926; November, 1926–May, 1927; May, 1927–November, 1927.

(with M. Heidelberger, W. F. Goebel, W. S. Tillett, L. A. Julianelle, M. H. Dawson, and E. G. Stillman)

Julianelle, L. A. A biological classification of *Encapsulatus pneumoniae* (Friedländer's bacillus). *J. Exp. Med.* 44:113.

Goebel, W. F. On the oxidation of glucose in alkaline solutions of iodine. *J. Biol. Chem.* 72:801.

Goebel, W. F. The preparation of hexonic and bionic acids by oxidation of aldoses with barium hypoiodite. *J. Biol. Chem.* 72:809.

Heidelberger, M. The chemical nature of immune substances. *Physiol. Rev.* 7:107.

Heidelberger, M. Immunologically specific polysaccharides. *Chem. Rev.* 3:423.

Heidelberger, M., Goebel, W. F. The soluble specific substance of pneumococcus. IV. On the nature of the specific polysaccharide of Type III pneumococcus. *J. Biol. Chem.* 70:613.

Julianelle, L. A. Immunological relationships of encapsulated and capsule-free strains of *Encapsulatus pneumoniae* (Friedländer's bacillus). *J. Exp. Med.* 44:683.

Julianelle, L. A. Immunological relationships of cell constituents of *Encapsulatus pneumoniae* (Friedländer's bacillus). *J. Exp. Med.* 44:735.

Julianelle, L. A., and Reimann, H. A. The production of purpura by derivatives of pneumococcus. III. Further studies on the nature of the purpura-producing principle. *J. Exp. Med.* 45:609.

Neill, J. M. Studies on the oxidation and reduction of immunological substances. V. Production of antihemotoxin by immunization with oxidized pneumococcus hemotoxin. *J. Exp. Med.* 45:105.

Stillman, E. G., and Branch, A. Susceptibility of rabbits to infection by the inhalation of virulent pneumococci. *J. Exp. Med.* 44:581–587.

Tillett, W. S. Studies on immunity to pneumococcus mucosus (Type III). I. Antibody response of rabbits immunized with Type III pneumococcus. *J. Exp. Med.* 45:713.

Stillman, E. G. The development of agglutinins and protective antibodies in rabbits following inhalation of pneumococci. *J. Exp. Med.* 45:1057.

1927–1928

(with W. S. Tillett, L. A. Julianelle, W. F. Goebel, R. J. Dubos, M. H. Dawson)

Tillett, W. S. Studies on immunity to pneumococcus mucosus. I. Antibody response of rabbits to Type III Pneumococcus. *J. Exp. Med.* 45:713.

Tillett, W. S. Studies on immunity to pneumococcus (Type III). II. The infectivity of Type III pneumococcus for rabbits. *J. Exp. Med.* 45:1093.

Tillett, W. S. Studies on immunity to pneumococcus mucosus (Type III). III. Increased resistance to Type III infection induced in rabbits by immunization with "R" and "S" forms of pneumococcus. *J. Exp. Med.* 46:343.

Goebel, W. F. The soluble specific substance of Friedländer's bacillus. IV. On the nature of the hydrolytic products of the specific carbohydrate from Type II. *J. Biol. Chem.* 74:619.

Heidelberger, M., and Goebel, W. F. The soluble specific substance of pneumococcus. V. On the chemical nature of the aldobionic acid from the specific polysaccharide of Type III pneumococcus. *J. Biol. Chem.* 74:613.

Goebel, W. F., and Avery, O. T. The soluble specific substance of Friedländer's bacillus. III. On the isolation and properties of the specific carbohydrates from Types A and C Friedländer's bacillus. *J. Exp. Med.* 46:601.

Julianelle, L. A., and Reimann, H. A. The production of purpura by derivatives of pneumococcus. III. Further studies on the nature of the purpura producing principle. *J. Exp. Med.* 45:609.

Dawson, M. H., and Avery, O. T. Reversion of avirulent "Rough" forms of pneumococcus to virulent "Smooth" types. *Proc. Soc. Exp. Biol. Med.* 24:943.

Goebel, W. F. The preparation of hexonic and bionic acids by oxidation of aldoses with barium hyporodite. *J. Biol. Chem.* 72:809.

Goebel, W. F. On the oxidation of glucose in alkaline solutions of iodine. *J. Biol. Chem.* 72:801.

1928–1929
(with E. G. Stillman, W. S. Tillett, L. A. Julianelle, W. F. Goebel, R. J. Dubos, M. H. Dawson, T. Francis, Jr.)

Avery, O. T., and Tillett, W. S. Anaphylaxis with the type-specific carbohydrates of pneumococcus. *J. Exp. Med.* 49:251.

Dawson, M. H. The interconvertibility of "R" and "S" forms of pneumococcus. *J. Exp. Med.* 47:577.

Dubos, R. J. Observations on the oxidation-reduction properties of sterile bacteriological media. *J. Exp. Med.* 49:507.

Dubos, R. J. The initiation of growth of certain facultative anaerobes as related to oxidation-reduction processes in the medium. *J. Exp. Med.* 49:559.

Dubos, R. J. The relation of the bacteriostatic action of certain dyes to oxidation-reduction processes. *J. Exp. Med.* 49:575.

Goebel, W. F., and Avery, O. T. A study of pneumococcus autolysis. *J. Exp. Med.* 49:267.

Julianelle, L. A. Bacterial variation in cultures of Friedländer's bacillus. *J. Exp. Med.* 47:889.

Julianelle, L. A., and Avery, O. T. Intracutaneous vaccination of rabbits with pneumococcus. I. Antibody response. II. Resistance to infection. III. Hypersensitiveness. *Proc. Soc. Exp. Biol. Med.* 26:224.

Tillett, W. S. Active and passive immunity to pneumococcus infection induced in rabbits by immunization with "R" pneumococci. *J. Exp. Med.* 48:791.

1929–1930
(with E. G. Stillman, W. S. Tillett, L. A. Julianelle, W. F. Goebel, R. J. Dubos, T. Francis, Jr., W. Kelley, F. H. Babers)

Avery, O. T., and Goebel, W. F. Chemo-immunological studies on conjugated carbohydrate-proteins. II. Immunological specificity of synthetic sugar-protein antigens. *J. Exp. Med.* 50:521.

Dawson, M. H. The transformation of pneumococcal types. I. The conversion of R forms of pneumococcus into S forms or the homologous type. *J. Exp. Med.* 51:99.

Dawson, M. H. The transformation of pneumococcal types. II. The interconvertibility of type-specific S pneumococci. *J. Exp. Med.* 51:123.

Dubos, R. J. The role of carbohydrates in biological oxidations and reductions. Experiments with pneumococcus. *J. Exp. Med.* 50:143.

Goebel, W. F., and Avery, O. T. Chemo-immunological studies on conjugated carbohydrate-proteins. I. The synthesis of p-aminophenol β-glucoside p-aminophenol β-galactoside, and their coupling with serum globulin. *J. Exp. Med.* 50:535.

Heidelberger, M., Avery, O. T., and Goebel, W. F. A soluble specific substance derived from gum arabic. *J. Exp. Med.* 49:847.

Julianelle, L. A. Reactions of rabbits to injections of pneumococci and their products. I. The antibody response. *J. Exp. Med.* 51:441.

II. Resistance to infection. *J. Exp. Med.* 51:449.

III. Reactions at the site of injection. *J. Exp. Med.* 51:463.

IV. The development of skin reactivity to derivatives of Pneumococcus. *J. Exp. Med.* 51:625.

V. The development of eye reactivity to derivatives of Pneumococcus. *J. Exp. Med.* 51:633.

VI. Hypersensitiveness to pneumococci and their products. *J. Exp. Med.* 51:643.

Stillman, E. G., and Branch, A. Early pulmonary lesions in partially immune alcoholized mice following inhalation of virulent pneumococci. *J. Exp. Med.* 51:275.

Tillett, W. S., Avery, O. T., and Goebel. W. F. Chemo-immunological studies on conjugated carbohydrate-proteins. III. Active and passive anaphylaxis with synthetic sugar-proteins. *J. Exp. Med.* 50:551.

Tillett, W. S., and Francis, T., Jr. Cutaneous reactions to the polysaccharides and proteins of pneumococcus in lobar pneumonia. *J. Exp. Med.* 50:687.

1930–1931

(with E. G. Stillman, W. F. Goebel, R. J. Dubos, T. Francis, Jr., W. Kelley, F. H. Babers, K. Goodner, J. L. Alloway)

Avery, O. T., and Dubos, R. J. The specific action of a bacterial enzyme on pneumococci of Type III. *Science* 72:151.

Babers, F. H., and Goebel, W. F. The molecular size of the Type III specific polysaccharide of Pneumococcus. *J. Biol. Chem.* 89:387.

Dubos, R. J. The bacteriostatic action of certain components of commercial peptones as affected by conditions of oxidation and reduction. *J. Exp. Med.* 52:331.

Francis, T., Jr., and Tillett, W. S. Cutaneous reactions in Pneumonia. The development of antibodies following the intradermal injection of type specific polysaccharide. *J. Exp. Med.* 52:573.

Goebel, W. F. The preparation of the type-specific polysaccharides of Pneumococcus. *J. Biol. Chem.* 89:395.

Julianelle, L. A. The distribution of Friedländer's bacilli of different types. *J. Exp. Med.* 52:539.

Stillman, E. G. The effect of the route of immunization on the immunity response to Pneumococcus Type I. *J. Exp. Med.* 51:721.

Stillman, E. G. Susceptibility of rabbits to infection by the inhalation of Type II pneumococci. *J. Exp. Med.* 52:215.

Stillman, E. G. Development of agglutinins and protective antibodies in rabbits, after inhalation of Type II pneumococci. *J. Exp. Med.* 52:225.

Tillett, W. S., and Francis, T., Jr. Serological reactions in pneumonia with a non-protein somatic fraction of Pneumococcus. *J. Exp. Med.* 53:561; *J. Clin. Invest.* 9:11.

Tillett, W. S., Goebel, W. F., and Avery, O. T. Chemical and immunological properties of a species-specific carbohydrate of pneumococci. *J. Exp. Med.* 52:895.

1931–1932

(with E. G. Stillman, W. F. Goebel, R. J. Dubos, T. Francis, Jr., F. H. Babers, K. Goodner, J. L. Alloway, E. Terrell)

Alloway, J. L. The transformation *in vitro* of R pneumococci into S forms of different specific types by the use of filtered pneumococcus extracts. *J. Exp. Med.* 55:91.

Avery, O. T., and Dubos, R. The protective action of a specific enzyme against Type III pneumococcus infection in mice. *J. Exp. Med.* 54:471.

Avery, O. T., and Dubos, R. The specific action of a bacterial enzyme on Type III pneumococci. *Trans. Assoc. Amer. Phys.* 46:216.

Avery, O. T., and Goebel, W. F. Chemo-immunological studies on conjugated carbohydrate proteins. V. The immunological specificity of an antigen prepared by combining the capsular polysaccharide of Type III Pneumococcus with foreign protein. *J. Exp. Med.* 54:437.

Dubos, R., and Avery, O. T. Decomposition of the capsular polysaccharide of Pneumococcus Type III by a bacterial enzyme. *J. Exp. Med.* 54:51.

Dubos, R. Factors affecting yield of specific enzyme in cultures of the bacillus decomposing the capsular polysaccharide of Type III Pneumococcus. *J. Exp. Med.* 55:377.

Finkle, P. Metabolism of S and R forms of Pneumococcus. *J. Exp. Med.* 53:661.

Francis, T., Jr., and Tillett, W. S. Cutaneous reactions in rabbits to the type-specific capsular polysaccharides of Pneumococcus. *J. Exp. Med.* 54:587.

Francis, T. Jr., and Tillett, W. S. The significance of the type-specific skin test in the serum treatment of pneumonia, Type I. *J. Clin. Invest.* 10:659.

Francis, T., Jr. The identity of the mechanisms of type-specific agglutinin and precipitin reactions with Pneumococcus. *J. Exp. Med.* 55:55.

Goebel, W. F., and Avery, O. T. Chemo-immunological studies on conjugated carbohydrate proteins. IV. The synthesis of the p-aminobenzyl ether of the soluble specific substance of Type III Pneumococcus and its coupling with protein. *J. Exp. Med.* 54:451.

Goodner, K. The development and localization of the dermal pneumococcic lesion in the rabbit. *J. Exp. Med.* 54:847.

Goodner, K., Dubos, R., and Avery, O. T. The action of a specific enzyme in Type III pneumococcus dermal infection in rabbits. *J. Exp. Med.* 55:393.

Rhoads, C. P., and Goodner, K. The pathology of experimental dermal pneumococcus infection in the rabbit. *J. Exp. Med.* 54:41.

Stillman, E. G. Duration of demonstrable antibodies in the serum of rabbits immunized with heat-killed Type I pneumococci. *J. Exp. Med.* 54:615.

Stillman, E. G., and Branch, A. Localization of pneumococci in the lungs of partially immunized mice following inhalation of pneumococci. *J. Exp. Med.* 54:623.

1932–1933
(with E. G. Stillman, W. F. Goebel, R. J. Dubos, T. Francis, Jr., F. H. Babers, K. Goodner, E. Terrell, E. Rogers)

Avery, O. T. The role of specific carbohydrates in Pneumococcus infection and immunity. *Annals Int. Med.* 6:1 (John Phillips Memorial Prize).

Avery, O. T., Goebel, W. F., and Babers, F. H. Chemoimmunological studies on conjugated carbohydrate-proteins: VII. Immunological specificity of antigens prepared by combining α and β glucosides of glucose with proteins. *J. Exp. Med.* 55:769.

Alloway, J. L. Further observations on the use of Pneumococcus extracts in effecting transformation of Type *in vitro*. *J. Exp. Med.* 57:265.

Baudisch, O., and Dubos, R. Über Katalasewirkung von Eisenverbindungen in Kulturmedien. *Biochem. Zeit.* 245:278.

Francis, T., Jr. The value of the skin test with type-specific capsular polysaccharide in the serum treatment of Type I Pneumococcus pneumonia. *J. Exp. Med.* 57:617.

Goebel, W. F., Babers, F. H., and Avery, O. T. Chemo-immunological studies on conjugated carbohydrate-proteins. VI. The synthesis of p-aminophenol α-glucoside and its coupling with protein. *J. Exp. Med.* 55:761.

Goodner, K. A test for the therapeutic value of anti-pneumococcus serum. (Presented before the American Public Health Service, Washington, D. C., October 24, 1932) *J. Immunol.* 25:199.

Goodner, K., and Dubos, R. Studies on the quantitative action of a specific enzyme in Type III Pneumococcus dermal infection in rabbits. *J. Exp. Med.* 56:521.

Julianelle, L. A., and Rhoads, C. P. Reactions of rabbits to intracutaneous injections of pneumococci and their products. VII. The relation of hypersensitiveness to lesions in the lungs of rabbits infected with pneumococci. *J. Exp. Med.* 55:797.

Kelley, W. H. The antipneumococcus properties of normal swine serum. *J. Exp. Med.* 55:877.

Stillman, Ernest G. Reaction of rabbits following inhalation of Type III pneumococci. *J. Infect. Dis.* 50:542.

1933–1934
(with T. Francis, Jr., T. J. Abernethy, E. G. Stillman, R. J. Dubos, W. F. Goebel, F. H. Babers, K. Goodner, E. Rogers, E. Terrell)

Avery, O. T. Chemo-immunologische Untersuchungen an Pneumokokken-infektion und Immunität. *Sonderd. Naturwissenschaft.* 21:777.

Goebel, W. F., and Babers, F. H. Derivatives of glucuronic acid. I. The preparation of glucuronic acid from glucuron and a comparison of their reducing values. *J. Biol. Chem.* 100:573.

Goebel, W. F., and Babers, F. H. Derivatives of glucuronic acid. II. The acetylation of glucuron. *J. Biol. Chem.* 100:734.

Goebel, W. F., and Babers, F. H. Derivatives of glucuronic acid. III. The synthesis of diacetylchloroglucuron. *J. Biol.Chem.* 101:173.

Avery, O. T., and Goebel, W. F. Chemo-immunological studies on the soluble specific substance of Pneumococcus. Isolation and properties of the acetyl polysaccharide of Pneumococcus Type I. *J. Exp. Med.* 58:731.

Goodner, K., and Stillman, E. G. The evaluation of active resistance to Pneumococcus infection in rabbits. *J. Exp. Med.* 58:183.

Goodner, K. The effect of Pneumococcus autolysates upon pneumococcus dermal infection in the rabbit. *J. Exp. Med.* 58:153.

1934–1935

(with T. J. Abernethy, F. H. Babers, B. Chow, R. J. Dubos, W. F. Goebel, K. Goodner, F. L. Horsfall, Jr., C. M. MacLeod, and E. G. Stillman)

Babers, F. H., and Goebel, W. F. The synthesis of the p-aminophenol β-glycosides of maltose, lactose, cellobiose and gentiobiose. *J. Biol. Chem.* 105:473.

Francis, T., Jr., and Terrell, E. E. Experimental Type III pneumococcus pneumonia in monkeys. I. Production and clinical course. *J. Exp. Med.* 59:609.

Francis, T., Jr., Terrell, E. E., Dubos, R., and Avery, O. T. Experimental Type III pneumococcus pneumonia in monkeys. II. Treatment with an enzyme which decomposes the specific capsular polysaccharide of Pneumococcus Type III. *J. Exp. Med.* 59:641.

Goebel, W. F., Babers, F. H., and Avery, O. T. Chemo-immunological studies on conjugated carbohydrate-proteins. VIII. The influence of the acetyl group on the specificity of hexoside-protein antigens. *J. Exp. Med.* 60:85.

Goebel, W. F., Avery, O. T., and Babers, F. H. Chemo-immunological studies on conjugated carbohydrate-proteins. IX. The specificity of antigens prepared by combining the p-aminophenol glycosides of disaccharides with protein. *J. Exp. Med.* 60:599.

Goebel, W. F., and Babers, F. H. Derivatives of glucuronic acid. IV. The synthesis of α and β tetracetyl glucuronic acid methyl ester and of l-chlorotriacetyl glucuronic acid methyl ester. *J. Biol. Chem.* 106:63.

Goodner, Kenneth. Studies on host factors in pneumococcus infections. I. Certain factors involved in the curative action of specific antipneumococcus serum in Type I pneumococcus dermal infection in rabbits. *J. Exp. Med.* 60:9–18.

Goodner, Kenneth. Studies on host factors in pneumococcus infections. II. The protective action of Type I antipneumococcus serum in rabbits. *J. Exp. Med.* 60:19.

Pittman, M., and Goodner, K. Complement-fixation with the type-specific carbohydrate of hemophilus influenzae. Type B. *J. Immunol.* 29:239.

Stillman, E. G. Duration of demonstrable antibodies in the serum of rabbits immunized with heat-killed Type II and Type III Pneumococci. *J. Infect. Dis.* 57:223.

Stillman, E. G., and Schulz, R. Z. The reaction of partially immunized rabbits to inhalation of Type I pneumococci. *J. Infect. Dis.* 57:233.

Stillman, E. G., and Schulz, R. Z. The reaction of normally and partially immunized rabbits to intranasal instillation of Type I pneumococci. *J. Infect. Dis.* 57:238.

1935–1936
(with T. J. Abernethy, R. J. Dubos, W. F. Goebel, K. Goodner, R. D. Hotchkiss,
F. L. Horsfall, Jr., C. M. MacLeod, and E. G. Stillman)

Abernethy, T. J., Horsfall, F. L., Jr., and MacLeod, C. M. Pneumothorax therapy in lobar pneumonia. *Bull. Johns Hopkins Hosp.* 58:35.

Dubos, R. Studies on the mechanism of production of a specific bacterial enzyme which decomposes the capsular polysaccharide of Type III Pneumococcus. *J. Exp. Med.* 62:259.

Dubos, R., and Bauer, J. H. The use of graded collodion membranes for the concentration of a bacterial enzyme capable of decomposing the capsular polysaccharide of Type III Pneumococcus. *J. Exp. Med.* 62:271.

Goodner, K., and Horsfall, F. L., Jr. The protective action of Type I antipneumococcus serum in mice. I. The quantitative aspects of the mouse protection test. *J. Exp. Med.* 62:359.

Goodner, K., and Miller, D. K. The protective action of Type I antipneumococcus serum in mice. II. The course of the infectious process. *J. Exp. Med.* 62:375.

Goodner, K., and Miller, D. K. The protective action of Type I antipneumococcus serum in mice. III. The significance of certain host factors. *J. Exp. Med.* 62:393.

Horsfall, F. L., Jr., and Goodner, K. Relation of the phospholipins to the reactivity of antipneumococcus sera. *Proc. Soc. Exp. Biol. Med.* 32:1329.

Swift, H. F., Lancefield, R. C., and Goodner, K. The serologic classification of hemolytic streptococci in relation to epidemiologic problems. *Am. J. Med. Sc.* 190:445.

Horsfall, F. L., Jr., and Goodner, K. Lipoids and immunological reactions. I. The relation of phospholipins to the type-specific reactions of antipneumococcus horse and rabbit sera. *J. Exp. Med.* 62:485.

Goebel, W. F., and Babers, F. H. Derivatives of glucuronic acid. V. The synthesis of glucuronides. *J. Biol. Chem.* 110:707.

Goebel, W. F., and Babers, F. H. Derivatives of glucuronic acid. VI. The preparation of α-chloro- and α-bromotriacetylglucuronic acid methyl ester, and the synthesis of β-glucuronides. *J. Biol. Chem.* 111:347.

Goebel, W. F. Chemo-immunological studies on the soluble specific substance of Pneumococcus. II. The chemical basis for the immunological relationship between the capsular polysaccharides of Types III and VIII Pneumococcus. *J. Biol. Chem.* 110:391.

Chow, B. F., and Goebel, W. F. The purification of the antibodies in Type I antipneumococcus serum, and the chemical nature of the type-specific precipitin reaction. *J. Exp. Med.* 62:179.

1936–1937
(with R. J. Dubos, W. F. Goebel, K. Goodner, F. L. Horsfall, Jr., R. D.
Hotchkiss, C. M. MacLeod, and E. G. Stillman)

Abernethy, T. J. Concentrated antipneumococcus serum in Type I pneumonia. Control of dosage by skin tests with type specific polysaccharide. *N. Y. State J. Med.* 36:627.

Abernethy, T. J., and Francis, T. Jr. Studies on the somatic "C" polysaccharide of pneumococcus I. Cutaneous and serological reactions in pneumonia. *J. Exp. Med.* 65:59.

Abernethy, T. J. Studies on the somatic "C" polysaccharide of pneumococcus II. The precipitation reaction in animals with experimentally induced pneumococcic infection. *J. Exp. Med.* 65:75.

Dubos, R., and Miller, B. Enzyme for decomposition of creatinine and its action on the "apparent creatinine" of blood. *Proc. Soc. Exp. Biol. Med.* 35:335.

Dubos, R., Meyer, K., and Smyth, E. M. Action of the lytic principle of pneumococcus on certain tissue polysaccharides. *Proc. Soc. Exp. Biol. Med.* 34:816.

Dubos, R., Meyer, K., and Smyth, E. M. The hydrolysis of the polysaccharide acid of vitreous humors, of umbilical cord, and of streptococcus by the autolytic enzyme of pneumococcus. *J. Biol. Chem.* 118:71.

Goebel, W. F. Chemo-immunological studies on conjugated carbohydrate-proteins. X. The immunological properties of an artificial antigen containing glucuronic acid. *J. Exp. Med.* 64:29.

Goebel, W. F., and Reznikoff, P. The preparation of ferrous gluconate and its use in the treatment of hypochromic anemia in rats. *J. Pharmacol. Exp. Ther.* 59:162.

Goodner, K., Horsfall, F. L., and Bauer, J. H. Ultrafiltration of Type I pneumococcal sera. *Proc. Soc. Exp. Biol. Med.* 34:617.

Goodner, K., and Horsfall, F. L. The protective action of Type I antipneumococcus serum in mice. IV. The prozone. V. The effect of added lipids on the protective mechanism. *J. Exp. Med.* 64:369 and 377.

Goodner, K., and Horsfall, F. L. The complement fixation reaction with pneumococcus capsular polysaccharide. *J. Exp. Med.* 64:201.

Horsfall, F. L., and Goodner, K. Lipids and immunological reactions. II. Further experiments on the relation of lipids to the type specific reactions of antipneumococcus sera. *J. Immunol.* 31:135.

Horsfall, F. L., and Goodner, K. Lipids and immunological reactions. III. Lipid content of specific precipitates from Type I antipneumococcus sera. *J. Exp. Med.* 64:583.

Horsfall, F. L., and Goodner, K. Lipids and immunological reactions. IV. The lipid patterns of specific precipitates from Type I antipneumococcus sera. *J. Exp. Med.* 64:855.

Horsfall, F. L., Goodner, K., and MacLeod, C. M. Type specific antipneumococcus rabbit serum. *Science* 84:579.

Hotchkiss, R. D., and Goebel, W. F. The synthesis of the aldobionic acid of gum acacia. *J. Am. Chem. Soc.* 58:858.

Hotchkiss, R. D., and Goebel, W. F. The synthesis of the heptacetyl methyl ester of gentiobiuronic acid. *Science* 83:353.

Hotchkiss, R. D., and Goebel, W. F. Derivatives of glucuronic acid. VII. The synthesis of aldobionic acids. *J. Biol. Chem.* 115:285.

1937–1938

(with P. B. Beeson, R. J. Dubos, W. F. Goebel, K. Goodner, C. L. Hoagland, F. L. Horsfall, Jr., R. D. Hotchkiss, C. M. MacLeod, R. Reeves, E. G. Stillman and R. H. S. Thompson)

Beeson, P. B., and Hoagland, C. L. Use of calcium chloride in relief of chills following serum administration. *Proc. Soc. Exp. Biol. Med.* 38:160.

Dubos, R. J. The decompositon of yeast nucleic acid by a heat resistant enzyme. *Science* 85:549.

Dubos, R. J., and MacLeod, C. M. Effect of a heat-resistant enzyme upon the antigenicity of pneumococci. *Proc. Soc. Exp. Biol. Med.* 36:696.

Dubos, R. J., and Miller, B. The production of bacterial enzymes capable of decomposing creatinine. *J. Biol. Chem.* 121:429.

Dubos, R. J. The effect of formaldehyde on pneumococci. *J. Exp. Med.* 67:389.

Goebel, W. F. The chemical constitution of benzoyl glucuronic acid. *Science* 86:105.

Goebel, W. F., and Hotchkiss, R. D. Chemo-immunological studies on conjugated carbohydrate-proteins. XI. The specificity of azo-protein antigens containing glucuronic and galacturonic acids. *J. Exp. Med.* 66:191.

Goebel, W. F., Reeves, R., and Hotchkiss, R. D. The synthesis of aldobionides. *J. Am. Chem. Soc.* 59:2745.

Goebel, W. F. Derivatives of glucuronic acid. VIII. The structure of benzoylglucuronic acid. *J. Biol. Chem.* 122:649.

Goodner, K., and Horsfall, F. L. Jr. Properties of the type specific proteins of antipneumococcus sera. I. The mouse protective value of Type I sera with reference to the precipitin content. *J. Exp. Med.* 413:66.

Goodner, K., and Horsfall, F. L. Jr. Properties of the type specific proteins of antipneumococcus sera. II. Immunological fractionation of Type I antipneumococcus horse and rabbit sera. *J. Exp. Med.* 66:425.

Goodner, K., and Horsfall, F. L. Jr. Properties of the type specific proteins of antipneumococcus sera. III. Immunochemical fractionation of Type I antipneumococcus horse and rabbit sera. *J. Exp. Med.* 66:437.

Goodner, K., and Horsfall, F. L. Jr. Passive anaphylactic sensitivity to pneumococcal capsular polysaccharides. *J. Immunol.* 33:259.

Goodner, K., Horsfall, F. L., Jr., and Dubos, R. Type specific antipneumococcic rabbit serum for therapeutic purposes. Production, processing and standardization. *J. Immunol.* 33:279.

Goodner, K., Horsfall, F. L., Jr. The purpuric reaction produced in animals by derivatives of the Pneumococcus. *Proc. Soc. Exp. Biol. Med.* 37:178.

Horsfall, F. L. Jr. The control of lobar pneumonia. *Can. J. Public Health, October,* 476.

Horsfall, F. L., Jr., Goodner, K., and MacLeod, C. M. Antipneumococcus rabbit serum as a therapeutic agent in lobar pneumonia. II. Additional observations in pneumococcus pneumonias of nine different types. *N. Y. State J. Med.* 38:1.

Hotchkiss, R. D., and Goebel, W. F. Chemo-immunological studies on the soluble specific substance of Pneumococcus III. The structure of the aldobionic acid from the Type III polysaccharide. *J. Biol. Chem.* 212:195.

MacLeod, C. M., and Farr, L. E. Relation of the carrier state to pneumococcal peritonitis in young children with the nephrotic syndrome. *Proc. Soc. Exp. Biol. Med.* 37:556.

Miller, B. F., and Dubos, R. Studies on the presence of creatinine in human blood. *J. Biol. Chem.* 121:448.

Miller, B. F., and Dubos, R. Determination by a specific, enzymatic method of the creatinine content of blood and urine from normal and nephritic individuals. *J. Biol. Chem.* 121:457.

Reznikoff, P., and Goebel, W. F. The use of ferrous gluconate in the treatment of hypochromic anemia. *J. Clin. Invest.* 16:547.

Stillman, E. G. The susceptibility of mice to inhaled Type III pneumococci. *J. Infect. Dis.* 62:66.

1938-1939

(with M. H. Adams, P. B. Beeson, C. Cattaneo, G. Daddi, A. W. Downie, T. Dublin, R. J. Dubos, W. F. Goebel, K. Goodner, R. Heggies, C. L. Hoagland, R. D. Hotchkiss, C. M. MacLeod, R. E. Reeves, and E. G. Stillman)

Downie, A. W. Antipneumococcus species immunity. *J. Hyg.* 38:292.

Downie, A. W. Antigenic activity of extracts of pneumococci. *J. Hyg.* 38:279.

Dubos, R. J. Immunization of experimental animals with a soluble antigen extracted from pneumococci. *J. Exp. Med.* 67:799.

Dubos, R. J. The bactericidal effect of an extract of a soil bacillus on Gram positive cocci. *Proc. Soc. Exp. Biol. Med.* 40:311.

Dubos, R. J., and MacLeod, C. M. The effect of a tissue enzyme upon pneumococci. *J. Exp. Med.* 67:791.

Dubos, R. J., and Miller, B. F. A bacterial enzyme which converts creatine into its anhydride creatinine. *Proc. Soc. Exp. Biol. Med.* 39:65.

Dubos, R. J., and Thompson, R. H. S. The decomposition of yeast nucleic acid by a heat-resistant enzyme. *J. Biol. Chem.* 124:501.

Goebel, W. F. The isolation of the blood group A specific substance from commercial peptone. *J. Exp. Med.* 68:221.

Goebel, W. F. Chemo-immunological studies on conjugated carbohydrate-proteins. XII. The immunological properties of an artificial antigen containing cellobiuronic acid. *J. Exp. Med.* 68:469.

Goebel, W. F. Immunity to experimental pneumococcus infection with an artificial antigen. *Nature* 143:77.

Goebel, W. F., and Reeves, R. E. Derivatives of glucuronic acid. IX. The synthesis of aldobionides and the relationship between the molecular rotation of derivatives of acetylated aldoses and uronic acids. *J. Biol. Chem.* 124:207.

Goodner, K., Horsfall, F. L. Jr., and Bauer, J. H. Some factors which affect the ultrafiltration of antipneumococcal sera. *J. Immunol.* 35:439.

Hoagland, C. L., Beeson, P. B., and Goebel, W. F. The capsular polysaccharide of the Type XIV Pneumococcus and its relationship to the specific substances of human blood. *Science* 88:261.

Horsfall, F. L. Jr. The characteristics of antipneumococcus sera produced by various animal species. *J. Bacteriol.* 35:207.

Lavin, G. I., Thompson, R. H. S., and Dubos, R. J. The ultraviolet absorption spectra of fractions isolated from pneumococci. *J. Biol. Chem.* 125:75.

MacLeod, C. M. Treatment of pneumonia with antipneumococcal rabbit serum. *Bull. N. Y. Acad. Med.* 15:116.

MacLeod, C. M., Hoagland, C. L., and Beeson, P. B. The use of the skin test with the type specific polysaccharides in the control of serum dosage in pneumococcal pneumonia. *J. Clin. Invest.* 17:739.

Stillman, E. G. Viability of pneumococci in dried sputum. *J. Infect. Dis.* 63:340.

Thompson, R. H. S., and Dubos, R. J. Production of experimental osteomyelitis in rabbits by intravenous injection of staphylococcus aureus. *J. Exp. Med.* 68:191.

Thompson, R. H. S., and Dubos, R. J. The isolation of nucleic acid and nucleoprotein fractions from pneumococci. *J. Biol. Chem.* 125:65.

1939–1940

(with M. H. Adams, P. B. Beeson, C. Cattaneo, E. C. Curnen, G. Daddi, T. Dublin, R. J. Dubos, W. F. Goebel, K. Goodner, T. Heggies, R. D. Hotchkiss, C. M. MacLeod, G. S. Mirick, and E. G. Stillman)

Beeson, P. B., and Goebel, W. F. The immunological relationship of the capsular polysaccharide of Type XIV Pneumococcus to the blood group A specific substance. *J. Exp. Med.* 70:239.

Dubos, R. J. Enzymatic analysis of the antigenic structure of pneumococci. *Ergebn. Enzymforsch.* 8:135.

Dubos, R. J. Studies on a bactericidal agent extracted from a soil bacillus. I. Preparation of the agent. Its activity *in vitro*. II. Protective effect of the bactericidal agent against experimental pneumococcus infections in mice. *J. Exp. Med.* 70:1 and 11.

Dubos, R. J., and Cattaneo, C. Studies on a bactericidal agent extracted from a soil bacillus. III. Preparation and activity of a protein-free fraction. *J. Exp. Med.* 70:249.

Goebel, W. F. Studies on antibacterial immunity induced by artificial antigens. I. Immunity to experimental pneumococcal infection with an antigen containing cellobiuronic acid. *J. Exp. Med.* 69:353.

Goebel, W. F. Immunity to experimental pneumococcal infection with an artificial antigen containing a saccharide of synthetic origin. *Science* 91:20.

Goebel, W. F., Beeson, P. B., and Hoagland, C. L. Chemo-immunological studies on the soluble specific substance of pneumococcus. IV. The capsular polysaccharide of Type XIV Pneumococcus and its relationship to the blood group A specific substance. *J. Biol. Chem.* 129:455.

Goodner, K., Horsfall, F. L. Jr., and Bauer, J. H. The neutralization of pneumococcal capsular polysaccharide by the antibodies of type-specific antisera. *J. Immunol.* 35:451.

Hotchkiss, R. D., and Dubos, R. J. Fractionation of the bactericidal agent from cultures of a soil bacillus. *J. Biol. Chem.* 132:791.

Hotchkiss, R. D., and Dubos, R. J. Chemical properties of bactericidal substances isolated from cultures of a soil bacillus. *J. Biol. Chem.* 132:793.

MacLeod, C. M. Chemotherapy of pneumococcic pneumonia. *J.A.M.A.* 113:1405.

MacLeod, C. M. Metabolism of "sulfapyridine-fast" and parent strains of Pneumococcus Type I. *Proc. Soc. Exp. Biol. Med.* 41:215.

MacLeod, C. M., and Daddi, G. A "sulfapyridine-fast" strain of Pneumococcus Type I. *Proc. Soc. Exp. Biol. Med.* 41:69.

Reeves, R. E. Saccharolactone methyl ester. *J. Am. Chem. Soc.* 61:664.

Smadel, J. E., Lavin, G. I., and Dubos, R. J. Some constituents of elementary bodies of vaccina. *J. Exp. Med.* 71:373.

Stillman, E. G., and Schulz, R. Z. Difference in virulence of various types of pneumococci for mice. *J. Infect. Dis.* 65:246.

1940–1941

(with M. H. Adams, A. Coburn, E. C. Curnen, R. J. Dubos, W. F.

Goebel, K. Goodner, R. D. Hotchkiss, C. M. MacLeod, G. S. Mirick, and E. G. Stillman)

Abernethy, T. J., and Avery, O. T. The occurrence during acute infections of a protein not normally present in the blood. I. Distrubution of the reactive protein in patients' sera and the effect of calcium on the flocculation reaction with C polysaccharide of Pneumococcus. *J. Exp. Med.* 73:173.

Beeson, P. B., and Goebel, W. F. Immunological cross-reactions of Type B Friedländer bacillus in Type II antipneumococcal horse and rabbit serum. *J. Immunol.* 38:231.

Beeson, P. B., and Hoagland, C. L. The use of calcium chloride in the treatment of chills. *N. Y. State J. Med.* 40:803.

Dubos, R. J. The effect of specific agents extracted from soil microorganisms upon experimental bacterial infections. *Ann. Int. Med.* 13:2025.

Dubos, R. J. The adaptive production of enzymes by bacteria. *Bacteriol. Rev.* 4:1.

Dubos, R. J. The utilization of selective microbial agents in the study of biological problems. *Harvey Lectures, Series 35,* 223.

Erf, L. A., and MacLeod, C. M. Increased urobilinogen excretion and acute hemolytic anemia in patients treated with sulfapyridine. *J. Clin. Invest.* 19:451.

Goebel, W. F. Studies on antibacterial immunity induced by artificial antigens. II. Immunity to experimental pneumococcal infection with antigens containing saccharides of synthetic origin. *J. Exp. Med.* 72:33.

Hotchkiss, R. D., and Dubos, R. J. Bactericidal fractions from an aerobic sporulating bacillus. *J. Biol. Chem.* 136:803.

Little, R. B., Dubos, R. J., and Hotchkiss, R. D. Action of gramicidin on streptococci of bovine mastitis. *Proc. Soc. Exp. Biol. Med.* 44:444.

Little, R. B., Dubos, R. J., and Hotchkiss, R. D. Effect of gramicidin suspended in mineral oil on streptococci of bovine mastitis. *Proc. Soc. Exp. Biol. Med.* 45:462.

Little, R. B., Dubos, R. J., and Hotchkiss, R. D. Gramicidin, Novoxil, and acriflavine for the treatment of the chronic form of streptococcic mastitis. *J. Am. Vet. Med. Assoc.* 98(No. 768):189.

MacLeod, C. M. The inhibition of the bacteriostatic action of sulfonamide drugs by substances of animal and bacterial origin. *J. Exp. Med.* 72:217.

MacLeod, C. M., and Avery, O. T. The occurrence during acute infections of a protein not normally present in the blood. II. Isolation and properties of the reactive protein. *J. Exp. Med.* 73:183.

MacLeod, C. M., and Avery, O. T. The occurrence during acute infections of a protein not normally present in the blood. III. Immunological properties of the C-reactive protein and its differentiation from normal blood proteins. *J. Exp. Med.* 73:191.

MacLeod, C. M., and Mirick, G. S. Bacteriological diagnosis of pneumonia in relation to chemotherapy. *Am. J. Public Health* 31:34.

MacLeod, C. M., Mirick, G. S., and Curnen, E. C. Toxicity for dogs of a bactericidal substance derived from a soil bacillus. *Proc. Soc. Exp. Biol. Med.* 43:461.

Reeves, R. E. The structure of trimethyl glucurone. *J. Am. Chem. Soc.* 4:1.

Reeves, R. E., Adams, M. H., and Goebel, W. F. The synthesis of a new dimethyl-β-methylglucoside. *J. Am. Chem. Soc.* 62:2881.

Stillman, E. G. The viability of pneumococci in dried rabbit blood. *J. Infect. Dis.* 66:171.

Stillman, E. G., and Schulz, R. Z. Susceptibility of mice to intranasal instillation of various types of pneumococci. *J. Infect. Dis.* 66:174.

1941–1942
(with M. H. Adams, E. C. Curnen, W. F. Goebel, F. L. Horsfall, Jr., M. McCarty, C. M. MacLeod, G. S. Mirick, and E. G. Stillman)

Adams, M. H., Reeves, R. E., and Goebel, W. F. The synthesis of 2,4-dimethyl-β-methyl-glucoside. *J. Biol. Chem.* 140:653.

Curnen, E. C., and MacLeod, C. M. The effect of sulfapyridine upon the development of immunity to pneumococcus in rabbits. *J. Exp. Med.* 75:77.

Horsfall, F. L., Jr. Recent studies in influenza. *Am. J. Public Health* 31:1275.

Reeves, R. E., and Goebel, W. F. Chemoimmunological studies on the soluble specific substance of pneumococcus. V. The structure of the Type III polysaccharide. *J. Biol. Chem.* 139:511.

Stillman, E. G. A comparison of antibody production by rabbits following injection of pneumococcus vaccines heated at 60°C. or autoclaved. *J. Immunol.* 41:343.

Stillman, E. G. The preservation of pneumococcus by freezing and drying. *J. Bacteriol.* 42:689.

1942–1943
(with M. H. Adams, O. Binkley, E. C. Curnen, W. F. Goebel, F. L. Horsfall, Jr., M. McCarty, G. S. Mirick, E. Perlman, E. G. Stillman, and J. E. Ziegler, Jr.)

Adams, M. H. The reaction between the enzyme tyrosinase and its specific antibody. *J. Exp. Med.* 76:175.

Horsfall, F. L. Jr. The present status of the influenza problem. *J. A. M. A.* 120:284.

Horsfall, F. L. Jr. *Human Influenza.* Cornell Univ. Press, New York, 1942.

Horsfall, F. L. Jr. The effect of sulphonamides on virulence of pneumococci. *J. Clin. Invest.* 21:647 (Abstr.).

Horsfall, F. L. Jr., Curnen, E. C., Mirick, G. S., Thomas, L., and Ziegler, J. E., Jr. A virus recovered from patients with primary atypical pneumonia. *Science* 97:289.

MacLeod, C. M., and McCarty, M. Relation of a somatic factor to virulence of pneumococci. *J. Clin. Invest.* 21:647 (Abstr.).

MacLeod, C. M., and Mirick, G. S. Quantitative determination of the bacteriostatic effect of the sulfonamide drugs on pneumococci. *J. Bacteriol.* 44:277.

Mirick, G. S. Mode of action of the sulphonamide drugs *in vitro. J. Bacteriol.* 45:66 (Abstr.).

Mirick, G. S. Enzymatic identification of p-amino benzoic acid (PAB) in cultures of pneumococcus and its relation to sulphonamide-fastness. *J. Clin. Invest.* 21:628 (Abstr.).

Thomas, L., Curnen, E. C., Mirick, G. S., Ziegler, J. E., Jr., and Horsfall, F. L., Jr. Complement fixation with dissimilar antigens in primary atypical pneumonia. *Proc. Soc. Exp. Biol. Med.* 52:121.

1943–1944
(with M. McCarty)

Avery, O. T. Karl Landsteiner. *J. Path. Bact.* 56:592.

Avery, O. T., MacLeod, C. M., and McCarty, M. Transformation of pneumococcal types induced by a desoxyribonucleic acid fraction isolated from Pneumococcus Type III. *J. Exp. Med.* 79:137.

1944–1945
(with M. McCarty)

McCarty, M. Reversible inactivation of the substance inducing transformation of pneumococcal types. *J. Clin. Invest.* 23:942 (Abstr.).

1945–1946
(with M. McCarty and H. F. Taylor)

Avery, O. T. Acceptance of the Kober Medal Award. *Trans. Assoc. Amer. Phys.* 59:43.

McCarty, M. Purification and properties of desoxyribonuclease isolated from beef pancreas. *J. Gen. Physiol.* 29:123.

McCarty, M., and Avery, O. T. Studies on the chemical nature of the substance inducing transformation of pneumococcal types. II. Effect of desoxyribonuclease on the biological activity of the transforming substance. *J. Exp. Med.* 83:89.

McCarty, M., and Avery, O. T. Studies on the chemical nature of the substance inducing transformation of pneumococcal types. III. An improved method for the isolation of the transforming substance and its application to Pneumococcus Types II, III and VI. *J. Exp. Med.* 83:97.

1946–1947
(with R. D. Hotchkiss, M. McCarty and H. F. Taylor)

McCarty, M. Chemical nature and biological specificity of the substance inducing transformation of pneumococcal types. *Bacteriol. Rev.* 10:63.

1947–1948
(with R. D. Hotchkiss, and H. F. Taylor)

Hotchkiss, R. D. The assimilation of amino acids by respiring washed staphylococci. *Fed. Proc.* 6:263.

Hotchkiss, R. D. A microchemical reaction resulting in the staining of polysaccharide structures in fixed tissue preparations. *Arch. Biochem.* 16:131.

Hotchkiss, R. D. Review of *Modern Development of Chemotherapy*, Havinga, E., et al., Amsterdam, 1946. *J. Appl. Physics* 18:1135.

McCarty, M., Taylor, H. F., and Avery, O. T. Biochemical studies of environmental factors essential in transformation of pneumococcal types. *Cold Spring Harbor Symp.* 11:177.

–1950

Avery, O. T., and Sprofkin, B. E. Studies on the bacteriolytic properties of *Streptomyces albus* and its action on hemolytic streptococci. Semi-Annual Progress Report, Department of Defense (DDRDB-3). (Unpublished report)

APPENDIXES

APPENDIX I

A Letter from Avery to His Brother Roy, Dated May 26, 1943

(This is part of a letter from Avery to his brother Dr. Roy C. Avery. The first pages of the handwritten text were written on May 13, 1943; they are not reproduced here because they deal with family affairs in relation to Avery's proposed move from New York to join his family in Nashville, Tennessee. In fact, the entire letter is an explanation of the postponement of the move. Avery had reached the [then] mandatory retirement age of 65 at The Rockefeller Institute for Medical Research and was to become Emeritus Member in July, 1943.

The second part of the letter, dated May 26, is here reproduced. Although it is commonly believed that it presents the first written record of the role of DNA as carrier of genetic information, this is not quite true. All the facts and hypotheses mentioned in the letter are reported at length in the annual report that was submitted to the Board of Scientific Directors in the early spring of 1943.

Along with much factual information, the letter contains many phrases that Avery commonly used in everyday conversations. For example, after describing some properties of the transforming substance he adds, "Sounds like a virus — maybe a gene. But with mechanisms I am not now concerned — One step at a time — and the first is, what is the chemical nature of the transforming principle? Someone else can work out the rest. Of course, the problem bristles with implications. . . . It's lots of fun to blow bubbles — but it's wiser to prick them yourself before someone else tries to. . . . It's hazardous to go off half cocked — and embarrassing to have to retract later."

The letter was terminated "long after midnight" and Avery apologizes for its deficiencies. "I'm so tired and sleepy I'm afraid I have not made this very clear. . . . Forgive this rambling epistle." In reality, the letter is far from rambling. Its technical parts are largely taken from the annual report written some two months earlier and are presented with precision and clarity. Even when writing to his brother, the Professor could not avoid playing one of his Red Seal Records! He also ended the letter with Dickens' phrase that he loved to use in the laboratory: "God bless us, one and all.")

Dr. Gasser and Dr. Rivers have been very kind and have insisted on my staying on, providing me an ample budget and technical assistance to carry on the problem that I've been studying. I've not published anything about it — indeed have discussed it only with a few — because I'm not yet convinced that we have (as yet) sufficient evidence. However, I did talk to Ernest [Dr. Ernest Goodpasture, Vanderbilt University Medical School] about it in Washington and I hope he has told you — for I have intended telling you first of all. I felt he should know because it bears directly on my coming eventually to Nashville.

It is the problem of the transformation of pneumococcal types. You will recall that Griffith, in London, some 15 years ago described a technique whereby he could change one specific type into another specific type through the intermediate R form. For example: Type II → R → Type III. This he accomplished by injecting mice with a large amount of *heat killed* Type III cells together with a small inoculum of a *living R culture* derived from *Type II*. He noted that not infrequently the mice so treated died and from their heart blood he recovered living, encapsulated Type III pneumococci. This he could accomplish only by the use of mice. He failed to obtain transformation when the *same* bacterial mixture was incubated in broth. Griffith's original observations were repeated and confirmed both in our Lab and abroad by Neufeld, and others. Then you remember Dawson with us reproduced the phenomenon *in vitro* by adding a dash of anti-R serum to the broth cultures. Later Alloway used *filtered extracts* prepared from Type III cells and in the absence of formed elements and cellular debris induced the R cultures derived from Type II to become typical encapsulated III pneumococcus. This you may remember involved several and repeated transfers in serum broth — often as many as 5–6 — before the change occurred. But it did occur and once the reaction was induced, thereafter without further addition of the inducing extract, the organisms continued to produce the Type III capsules; that is the change was hereditary and transmissible in series in plain broth thereafter. For the past two years, first with MacLeod and now with Dr. McCarty I have been trying to find out what is the chemical nature of the substance in the bacterial extracts which induces this specific change. The crude extract (Type III) is full of capsular polysaccharide, C (somatic) carbohydrate, nucleoproteins, free nucleic acids of both the yeast and thymus type, lipids and other cell constituents. Try to find in that complex mixture the active principle!! Try to isolate and chemically identify the particular substance that will by itself when brought into contact with the R cell derived from Type II cause it to elaborate Type III capsular polysaccharide, and to acquire all the aristocratic distinctions of the same specific type of cells as that from which the extract was prepared! Some job — and full of heartaches and heart breaks. But at last *perhaps* we have it. The active substance is *not* digested by crystalline trypsin or chymotrypsin — It does not lose activity when treated with crystalline Ribonuclease which specifically breaks down yeast nucleic acid. The Type III capsular polysaccharide can be removed by digestion with the specific Type III enzyme without loss of transforming activity of a potent extract. The lipids can be extracted from such extracts by alcohol and ether at −12° C without impairing biological activity. The extract can be de-proteinized by Sevag Method (shaking with chloroform and amyl alcohol) until protein free and biuret negative. When extracts treated and purified to this extent, but still containing traces of protein, lots of C carbohydrate and nucleic acids of both the yeast and thymus types are further fractionated by the dropwise addition of absolute ethyl alcohol, an interesting thing occurs. When alcohol reaches a concentration of about 9/10 volume there separates out a fibrous substance which on stirring the mixture wraps itself about the glass rod like thread on a spool — and the other impurities stay behind as granular precipitate. The fibrous material is redissolved and the process repeated several times — In short, the substance is highly reactive and on elementary analysis conforms *very* closely to the theoretical values of pure *desoxyribose nucleic acid* (thymus type). Who could have guessed it? This type of nucleic acid has not to my knowledge been recognized in pneumococcus before — though it has been found in other bacteria.

Of a number of crude enzyme preparations from rabbit bone, swine kidney, dog intestinal mucosa, and *pneumococci,* and fresh blood serum of human, dog and rabbit, only those containing active depolymerase capable of breaking down known authentic samples of desoxyribose nucleic acid have been found to destroy the activity of our substance — indirect evidence but suggestive that the transforming principle as isolated may belong to this class of chemical substance. We have isolated highly purified substance of which as little as 0.02 of a *micro*gram is active in inducing transformation. In the reaction mixture (culture medium) this represents a dilution of 1 part in a hundred million — potent stuff that — and highly specific. This does not leave much room for impurities — but the evidence is not good enough yet. In dilution of 1:1000 the substance is highly *viscous* as an authentic preparation of desoxyribose nucleic acid derived from fish sperm. Preliminary studies with the ultracentrifuge indicate a molecular weight of approximately 500,000 — a highly polymerized substance.

We are now planning to prepare new batch and get further evidence of purity and homogeneity by use of ultracentrifuge and electrophoresis. This will keep me here for a while longer. If things go well I hope to go up to Deer Isle, rest awhile — Come back refreshed and try to pick up loose ends in the problem and write up the work. If we are right, and of course that's not yet proven, then it means that nucleic acids are *not* merely structurally important but functionally active substances in determining the biochemical activities and specific characteristics of cells — and that by means of a known chemical substance it is possible to induce *predictable* and *hereditary* changes in cells. This is something that has long been the dream of geneticists. The mutations they induce by X ray and ultraviolet are always unpredictable, random, and chance changes. If we are proven to be right — and of course that's a big *if* — then it means that both the chemical nature of the *inducing stimulus* is known and the chemical structure of the *substance produced* is also known — the former being thymus nucleic acid — the latter Type III polysaccharide. And both are thereafter reduplicated in the daughter cells and after innumerable transfers and without further addition of the inducing agent, the same active and specific transforming substance can be recovered far in excess of the amount originally used to induce the reaction. Sounds like a virus — may be a gene. But with mechanisms I am not now concerned — One step at a time — and the first is, what is the chemical nature of the transforming principle? Someone else can work out the rest. Of course, the problem bristles with implications. It touches the biochemistry of the thymus type of nucleic acids which are known to constitute the major part of the chromosomes but have been thought to be alike regardless of origin and species. It touches genetics, enzyme chemistry, cell metabolism and carbohydrate synthesis, etc. today it takes a lot of well documented evidence to convince anyone that the sodium salt of desoxyribose nucleic acid, protein-free, could possibly be endowed with such biologically active and specific properties and this evidence we are now trying to get. It's lots of fun to blow bubbles — but it's wiser to prick them yourself before someone else tries to. So there's the story Roy — right or wrong it's been good fun and lots of work. This supplemented by war work and general supervision of other important problems in the Lab has kept me busy, as you can well understand. Talk it over with Goodpasture but don't shout it around — until we're quite sure or at least as sure as present method permits. It's hazardous to go off half cocked — and embarrassing to have to retract later.

I'm so tired and sleepy I'm afraid I have not made this very clear. But I want you to know — and sure you will see that I cannot well leave this problem until we've got convincing evidence. Then I look forward and hope we may all be together — God and the war permitting — and living out our days in peace. What a lovely picture of dear Margaret. How is she and Cath — wish we could all meet in Deer Isle. I know Minnie has kept you all posted. Things go well with us despite this cruel war but Victory must come and I'm optimistic enough to look forward to happier days even if they are not perfect — We can take it — and still be happy.

Forgive this rambling epistle — with it goes my love and thought and hope of better things ahead —

"With heaps and heaps of love"
Affectionately and faithfully,
OTA

[A P.S. but not so designated]

If the Board in the Surgeon General's office meets at Camp Bragg as I think they may later on I shall take the opportunity of running over to Nashville for I want to talk over future plans and possibilities with you and Catherine. Do write if just a line — I want to know your reaction and don't hestitate to talk to Ernest — he knows it all and we talked it over very frankly.

Good night — it's long after mid-night and I have a busy day ahead. God bless us, one and all. Sleepy, well and happy —

APPENDIX II

Typescript Used by Avery when He Delivered His Address "The Commonwealth of Science" in May, 1941, as President of the Society of American Bacteriologists

-12-

of the spirit of inquiry. They are to the Commonwealth of Science what the Bill

of Rights is to the life of democracy.

On that occasion Sir Richard Gregory, President of the British Association

presented the following charter of science which was unanimously adopted by the

Conference. Its seven articles represent not a creed but a policy; they possess no

sanctity or finality, but they do represent the spirit of science.

DECLARATION OF SCIENTIFIC PRINCIPLES

1. Liberty to learn, opportunity to teach and power to understand are necessary for

    the extension of knowledge and we, as men of science, maintain that they cannot

    be sacrificed without degradation to human life.

2. Communities depend for their existence, their survival and advancement, on

    knowledge of themselves, and of the properties of things in the world around them.

3. All nations and all classes of society have contributed to the knowledge and

    utilization of natural resources, and to the understanding of the influence they

    exercise on human development.

4. The service of science requires independence combined with cooperation and its

    structure is influenced by the progressive needs of humanity.

5. Men of science are among the trustees of each generation's inheritance of natural

    knowledge. They are bound, therefore, to foster and increase that heritage

APPENDIX III

The Lore of the Pneumococcus, as Presented by Avery
in an Annual Report to the Board of Trustees of The Rockefeller
Institute for Medical Research

(This text is taken from an essay prepared by Avery in the spring of 1931 as a supplement to the more technical annual report submitted at the same time to the Board of Scientific Directors of the Institute [see Appendix IV]. The essay was probably written at the request of Drs. Simon Flexner and Rufus Cole for the lay members of the Board of Trustees of the Institute. It presents in general terms Avery's views of the role of the capsular polysaccharides in the virulence of pneumococci. It also provides the background for the efforts to control type III pneumococcal infections by vaccination with a synthetic antigen. Later in the text, but not reproduced here, it discusses treatment with an enzyme capable of hydrolyzing the type III capsular polysaccharide.

Although the essay was written as part of the annual report for 1931, it can be read as an expression of the so-called "Red Seal Records," through which Avery presented the lore of the pneumococcus to his colleagues and to visitors.)

It has long been recognized that simple sugars such as glucose do not possess the property of an antigen, that is, they are incapable of stimulating the formation of antibodies in the animal body. However, it is now known that if these sugars are combined by chemical means with a protein, that is, with a substance naturally endowed with antigenic properties, the new sugar compounds thus formed incite the formation of antibodies that are specific for the particular sugar used. The study of synthetic antigens prepared by combining a simple sugar with protein has shown that the specificity of the newly formed compounds is determined by the chemical individuality of the reactive carbohydrate irrespective of the protein to which it is attached. Antisera produced by immunization with these conjugated sugar-proteins invariably reflect the controlling influence of the carbohydrate on the specificity of the whole compound. The studies on the simple non-bacterial sugars emphasize again the significance of carbohydrates in orienting the specific immune response of the body to substances of this class.

The results of this work led us to test the possibility of synthesizing an artificial pneumococcus antigen by combining the capsular polysaccharide with a foreign protein. For this purpose the polysaccharide of Type III was chosen since in its purified form it contains no nitrogen and represents a definite chemical entity. Further, if results were obtained with this particular sugar, they would be all the more interesting since the isolated pure substance itself has never been found to elicit antibodies in rabbits and even the intact cells from which it is derived frequently fail to incite antibody formation in these animals. By an intricate chemical synthesis, the details of which need not concern us here, it was found possible to combine the Type III capsular polysaccharide in stable chemical union with an unrelated protein of animal origin. This artificial antigen has in common with Type III Pneumococcus only the capsular polysaccharide, the protein with which it was combined being of widely remote biological origin. Rabbits injected with this artificial antigen were actively immune to subsequent infection and their serum specifically agglutinated living cultures of Type III Pneumococci, precipitated solutions of Type III polysaccharide, and protected mice against Type III Pneumococcus infection. In other words, this synthesized compound, containing only a single component of the pneumococcus cell, called forth an immune response as specific in nature as that induced by the whole microorganism.

APPENDIX IV

The Problems under Investigation in
Avery's Department during the 1930s

(The two documents presented here are the outlines of the annual reports submitted by Avery to the Board of Scientific Directors of the Institute in the springs of 1931 and 1937. These outlines, which are typical of those for the other years of the 1930s, give an idea of the wide range of theoretical and practical problems under investigation in the department during that period.

The 1931 report presents the first attempts to achieve the transformation of pneumococcal types with cell-free extracts of pneumococci. The 1937 report discusses some physicochemical properties of the transforming substance and, in particular, its susceptibility to inactivation by certain enzymes.)

Report of Dr. Avery with Drs. Stillman, Goebel, Dubos, Francis, Kelley, Babers, Goodner, and Alloway [1930–31]

   I. The decomposition of the Capsular Polysaccharide of Type III Pneumococcus by a Bacterial Enzyme.
      1. Methods for obtaining potent, non-toxic preparations of the enzyme.
      2. The protective action of the specific enzyme against Type III Pneumococcus infection in mice.
   II. Isolation of other microorganisms decomposing the capsular polysaccharides of different types of Pneumococcus.
   III. Chemo-immunological studies on Carbohydrates.
      1. The determination of the molecular size of the capsular polysaccharide of Type III Pneumococcus.
      2. The specific carbohydrate of two strains of Pfeiffer's bacillus.
      3. The somatic carbohydrate of Pneumococcus.
      4. Studies on synthetic carbohydrate derivatives.
         a) Synthesis of α para-amino phenol glucoside.
         b) Glucuronic acid.
   IV. Chemical Nature of Type Specific, Capsular antigen of Pneumococcus.
   V. Significance of the Skin Test as a guide to Serum Therapy in Pneumonia.
   VI. Studies on Natural Resistance and the Immunity induced by R pneumococci.
   VII. Principles Governing the Precipitin and Agglutinin reactions with Pneumococcus.

Report of Drs. Cole and Avery (assisted by Drs. Dubos, Goebel, Goodner, Horsfall, Hotchkiss, MacLeod and Stillman) [1936–37]

APPENDIX V

The Theory of Antigenic Dissociation

(Avery devoted the entire annual report that he submitted in the spring of 1927 to the analysis of the problems of virulence and antigenicity. He formulated in particular the hypothetical concept which he called "antigenic dissociation" – namely, the set of phenomena that, either *in vitro* or *in vivo,* bring about the separation of the capsular polysaccharide from the complex cellular structure of which it is a part in the pneumococcal cell. According to this hypothesis, antigenic dissociation is in some way associated with loss of antigenicity and loss of virulence.

The following is the table of contents of the section of the report dealing with antigenic dissociation. Although the hypothesis was never fully substantiated, it provided the framework for much of the immunological program carried out in Avery's department during the late 1920s and through the 1930s.)

Studies Concerning Chemistry and Immunological Properties of Pneumococcus. Report of Dr. Avery, with Drs. Heidelberger, Goebel, Tillett, Julianelle, and Dawson

STUDIES ON ANTIGENIC DISSOCIATION:

1. Pneumococcus as "Complex Antigen"
2. Consideration of the Cell as
   a. Two distinct and Separate antigenic systems.
   b. Single antigenic complex composed of carbohydrate (haptene) and protein.
3. Evidence for Antigenic Dissociation in Vitro:
   a. Qualitatively different antibodies stimulated by intact and dissolved cells.
   b. Relative differences in the dissociation of Types I, II, and III.
4. Evidence for Antigenic Dissociation in Vivo:
   a. Dissociation of Type III in rabbits.
   b. Antiprotein antibodies in serum as index of dissociation.
   c. Relation of antigenic dissociation to production of antipneumococcus sera.
   d. Antigenic potency inversely proportional to rate and extent of dissociation.
5. Factors relating to the Animal:
   a. Natural resistance and antibody response.
   b. Nature of bacterial injury.
   c. Difference between Natural Resistance and Specific Immunity.
6. Factors Relating to the Micro-organism:
   a. Relation between antigenic stability and chemical structure of the cell.
7. Concept of Virulence:
   a. As tissue fastness.
   b. Relationship between Virulence and Antigenicity.

APPENDIX VI

From Bacterial Transformation to Genetic Engineering

After Griffith's initial discovery, the transformation phenomenon acquired a life of its own within the walls of The Rockefeller Institute. There, DNA emerged as the transforming substance in Avery's laboratory, not through a planned, detailed, experimental program based on prevailing genetic theories, but as the result of a disciplined trial and error approach, a persistent day-to-day, step-by-step work at the bench, seemingly unaffected by outside influences.

This does not mean that the studies on transformation proceeded in an intellectual vacuum. Although Avery's laboratory was the only place where research was conducted on the subject until the late 1940s, many investigators in other institutions were involved during that period in problems that appeared unrelated to transformation, but eventually proved to have a direct bearing on the interpretation of the phenomenon and on its general significance. I shall mention a few of these problems in the following pages, not to present a documentation of the studies that preceded and followed the discovery that DNA is responsible for genetic specificity, but to evoke the intellectual atmosphere in which the discovery was made. The references that I have listed are not inclusive; they have been selected merely to illustrate the wide range of biological and chemical studies which, during the 1940s, created a scientific climate favorable to the conversion of such a crude biological phenomenon as pneumococcus type transformation in the mouse into the highly sophisticated science of molecular genetics.

Within a few years after the demonstration that pneumococci can be made to incorporate DNA from a different immunological type and thus undergo hereditary change in the chemical structure of their capsular polysaccharide, other phenomena of transformation were recognized in several other bacterial species. In little more than a decade, it had been shown that transformation can involve a great diversity of characters completely different from those that govern the synthesis of the capsular polysaccharides — for example, cellular morphology, production of many different enzymes, resistance to antibacterial agents, virulence for certain animal hosts. In brief, the transformation of immunological type in pneumococci proved to be but a special case of a very general phenomenon in the bacterial world. The rapidity with which the field developed is illustrated by the huge size of the bibliography in a review, "The Genetics of Transformation," that was published in 1961, 15 years after the publication of the original findings on the genetic role of DNA (reviewed in reference 1 to this Appendix).

Although many bacterial species have been shown to be capable of incorporating foreign DNA, recognition of the phenomenon in the laboratory is often made difficult by the fact that the recipient cells must be in a physiologically "competent" state before incorporation can take place; this state of "competence" exists for only a short period of the cell's growth cycle. Moreover, the ease of transformation is conditioned by a multiplicity of subtle environmental factors which differ from species to species (reviewed in reference 2). These experimental difficulties, however, are of minor importance when compared with the much larger conceptual difficulty posed by the chemical nature of the transforming substance.

The general view among geneticists then was that proteins are the only substances with a sufficient degree of chemical complexity to account for the immense diversity of hereditary processes. This view was challenged by the claim that hereditary characteristics can be transferred by molecules of deoxyribonucleic acid, but such a claim was chemically untenable as long as DNA was believed to be built up from simple tetranucleotides, which could not possibly provide the chemical diversity required for biological specificity. Next to the discovery of the transformation phenomenon itself, and its identification with DNA, the most fundamental step in the development of chemical genetics was therefore the demonstration that the molecular structure of DNA differs according to the biological source of the material.[3] It is of interest that E. Chargaff, whose role was crucial in discrediting the tetranucleotide doctrine, has repeatedly acknowledged Avery's influence on his own work:

"In 1944 Avery, MacLeod and McCarty published their famous paper on the transforming principle of pneumococci. This was really the decisive influence, as far as I was concerned, to devote our laboratory almost completely to the chemistry of nucleic acids. . . ."[4]

Once it had been shown, largely through Chargaff's work, that the distribution of bases differs from one biological type of DNA to another, it became possible to envisage chemical mechanisms for the specificity of genetic information, determined by the sequence of the four nucleotides along the polynucleotide chain.

Several other lines of studies which came to fruition during the 1940s made it possible to visualize how DNA functions as bearer of hereditary characteristics in the bacterial cell. New histological techniques were used to establish beyond doubt the existence of a nucleoid body in bacteria and to show that the cellular DNA is localized in this very structure.[5-7] Further evidence, even though indirect, for the genetic role of DNA was provided by two different kinds of findings, namely: the demonstration that this substance is the active infectious component of coli T-phage[8] and that diploid cells contain twice as much DNA as haploid cells.[9, 10]

During the 1940s also, it became increasingly apparent that the fundamental genetic mechanisms of bacteria are very similar to those in the cells of higher organisms. First came the demonstration that when hereditary changes take place in pure cultures of bacteria, they occur as discrete spontaneous events, resembling the mutations of classical genetics.[11] The relation between genes and transformation became even clearer in 1951, when it was shown that the very steps experienced in the mutational history of a bacterium could be faithfully recapitulated as discrete stages of transformation which its DNA could bring about. By this time, the discovery had been made that mating (conjugation) does occasionally occur in bacterial cells and results in exchange of genetic material between mated cells.[12] A few years later, it was found that genetic material can be introduced into bacteria through the agency of temperate phage, by a mechanism that has come to be known as transduction.[13,14] Conjugation, in particular, eventually revealed that bacteria possess chromosomelike structures which are linear arrays of many genes—a fact compatible with the view that the process of transformation in pneumococci is not really a mutation, but rather involves transfer of chromosomal material from donor to recipient cells.[15]

Transformation, conjugation, and transduction are thus three different mechanisms through which new genetic material can be introduced into the bacterial cell. To the extent that these mechanisms operate in nature—an extent which has not been determined quantitatively—they increase genetic variety by facilitating recombination of spontaneously occurring mutants. In this manner, they play a role in haploid, asexually reproducing bacteria similar to that of genetic recombination between chromosomes in diploid, sexually reproducing organisms. By increasing the flow of genes through populations of bacteria and probably between different species, transformation, conjugation, and transduction tend to produce a continuous spectrum of genetic differences, and thereby to blur species distinctiveness. If the flow of genes through these mechanisms were to occur widely on a large scale, the nineteenth-century doctrine of bacterial polymorphism might be reappearing under a new guise, but this is obviously not the case.[16]

For practical reasons, bacteriologists have found it useful to classify bacteria according to the Linnean system on the assumption that genetic discontinuity in these organisms is as marked as it is among higher organisms. Bergey's *Manual of Determinative Bacteriology* could not have been prepared without this assumption. There are many cases, of course, in which the conventional classification of bacteria in neatly defined genera and species appears to be more an exercise in textbook taxonomy than a reflection of the manner in which these organisms are related in nature. Nevertheless, practice shows that a number of bacterial groups are sufficiently stable and distinct from each other to make possible a fairly dependable system of classification. Such genetic discontinuity implies, of course, that there exist in bacteria certain mechanisms which act as barriers to gene flow and which thus play the same role as do genetic isolating mechanisms in higher organisms. Little is known concerning the mechanisms responsible for such interference to gene flow, but granted that genetic transfer between bacteria can be achieved by laboratory procedures, there is no evidence that the phenomenon occurs frequently in nature.

The very knowledge that foreign genetic material can be incorporated into a bacterial cell and that the genetic endowment can thus be modified has naturally encouraged investigators to pursue the problem beyond the limits of naturally occurring processes. Despite public alarm about the potential dangers of manipulating the genetic endowment of bacteria and other cells, it can be taken for granted that the biological methods of transformation, conjugation, and transduction will be used to "synthesize" composite cells possessing desired associations of properties. Such "chimera" are even now being produced in the laboratory, and will certainly continue to be produced. Furthermore, chemical manipulation of the DNA molecule may permit modifications of the genes themselves and thus lead to directed chemical mutation, for good or evil.

Genetic engineering, which is the popular name of this science, thus appears superficially to be a man-made reincarnation of bacterial polymorphism, one century after Louis Pasteur, Ferdinand Cohn, and Robert Koch had affirmed that bacterial species are stable and cannot change one into the other. The potentialities of genetic engineering are immense, and it is certain that, with all the genetic material of the world's biotype available, many creatures will be built that genetic isolation mechanisms would prevent from occurring in nature. But I doubt that, outside the laboratory, such genetic chimera will be a match for the products of bacterial variability as they occur under natural conditions. As I terminate this essay — at the end of my professional life — I experience as deeply as ever a feeling of humility before the inventiveness of nature, but also a sense of pride at man's ability to manipulate nature.

INDEX